Agro-Technology

Humans have been modifying plants and animals for millennia. The dawn of molecular genetics, however, has kindled intense public scrutiny and controversy. Crops, and the food products which include them, have dominated molecular modification in agriculture. Organisations have made unsubstantiated claims and scaremongering is common. In this textbook R. Paul Thompson presents a clear account of the significant issues – identifying harms and benefits, analysing and managing risk – which lie beneath the cacophony of public controversy. His comprehensive analysis looks especially at genetically modified organisms, and includes an explanation of the scientific background, an analysis of ideological objections, a discussion of legal and ethical concerns, a suggested alternative – organic agriculture – and an examination of the controversy's impact on sub-Saharan African countries. His book will be of interest to students and other readers in philosophy, biology, biotechnology and public policy.

R. PAUL THOMPSON is Professor at the Institute for the History and Philosophy of Science and Technology, and the Department of Ecology and Evolutionary Biology, at the University of Toronto.

Other titles in the Cambridge Introductions to Philosophy and Biology series:

Derek Turner, *Paleontology* 9780521116374

R. Paul Thompson, *Agro-Technology* 9780521117975

Agro-Technology

A Philosophical Introduction

R. PAUL THOMPSON

Institute for the History and Philosophy of Science and Technology, and the Department of Ecology and Evolutionary Biology, University of Toronto

CAMBRIDGE UNIVERSITY PRESS
Cambridge, New York, Melbourne, Madrid, Cape Town,
Singapore, São Paulo, Delhi, Tokyo, Mexico City

Cambridge University Press
The Edinburgh Building, Cambridge CB2 8RU, UK

Published in the United States of America by Cambridge University Press, New York

www.cambridge.org
Information on this title: www.cambridge.org/9780521117975

© R. Paul Thompson 2011

This publication is in copyright. Subject to statutory exception
and to the provisions of relevant collective licensing agreements,
no reproduction of any part may take place without the written
permission of Cambridge University Press.

First published 2011

A catalogue record for this publication is available from the British Library

Library of Congress Cataloguing in Publication data
Thompson, R. Paul, 1947–
Agro-technology : a philosophical introduction / R. Paul Thompson.
p. cm. – (Cambridge introductions to philosophy and biology)
Includes bibliographical references and index.
ISBN 978-0-521-11797-5 (hardback)
1. Agricultural biotechnology – Philosophy. 2. Genetic engineering –
Philosophy. 3. Agricultural biotechnology – Moral and ethical aspects.
4. Genetic engineering – Moral and ethical aspects. 5. Agricultural
biotechnology – Social aspects. 6. Genetic engineering – Social aspects. I. Title.
S494.5.B563T46 2011
630 – dc23 2011017974

ISBN 978-0-521-11797-5 Hardback
ISBN 978-0-521-13375-3 Paperback

Cambridge University Press has no responsibility for the persistence or
accuracy of URLs for external or third-party internet websites referred to
in this publication, and does not guarantee that any content on such
websites is, or will remain, accurate or appropriate.

For Olivia

May the wind be always at your back.

Contents

List of figures

List of tables

Preface

I have a long-standing personal interest in food: its history, its biology and chemistry, its production and its preparation. Hence, cooking provides a creative outlet, one in which my academic curiosity about the history, biology and chemistry of food can be combined with creating new methods of preparation, new ingredients and combinations of ingredients, and new combinations of flavours. Pursuing this interest has led me to delve into the history of food, especially the last 10–15, 000 years of the domestication of plants and animals and the introduction of novel foods in diverse regions of the globe, including wild sources of ingredients (see Elias and Dykeman, 1990; Gardon, 1998; Henderson, 2000; Thayer, 2006). It also has led me to study food chemistry and the cell and molecular properties of food, the transformation of food during preparation (such as the Maillard reaction when food is heated), the physiology and neuroscience of taste, and modern agricultural practices, food processing and food distribution. This book focuses mostly on the latter, specifically on biotechnology in agriculture and the controversy surrounding it.

I bring to the material in this book a special, though far from unique, combination of perspectives and knowledge. My academic interests breach the normal divide between science and the humanities. On the one side, I have a background in philosophy, hold an appointment in the Institute for the History and Philosophy of Science and Technology, and teach courses on the philosophy of biology and the philosophy of medicine. On the other, I also have a background in biology, hold an appointment in the Department of Ecology and Evolutionary Biology, and currently teach a biology course on molecular genetics and biotechnology. Over the last 30 years, I have taught biology courses on population genetics, evolution and epidemiology, and a diverse array of philosophy courses, including ethics, social issues, the philosophy of science, the philosophy of medicine and mathematical (symbolic) logic. I hope in the course of this book I can help others bridge what is often a deep chasm.

This is not an advocacy book but no one writes about issues as contentious as agricultural biotechnology without numerous influences, and preformed ideas and positions (hopefully positions based on the best available evidence and sound reasoning). Intellectual openness does not require coming to an issue with a blank slate or pretending to be positionless, but it does require that positions be open to change in the light of revised or new evidence, or exposed deficiencies in reasoning. To do otherwise is dogmatic and irrational.

A simple statement of thanks at the end of a preface dramatically underestimates the contribution made by so many to the ideas and analyses in this book. Some are long deceased philosophers reaching back to Plato and Aristotle. Others are contemporary researchers and scholars, from biologists to political scientists and economists to philosophers. Yet others are friends and colleagues. My long-standing and very close friends Michael Ruse and Paul Gooch opened up the rich and deeply important world of philosophical ideas and analysis. Hugh Grant, Jerry Steiner, Rob Horsh, Kate Fish and Dianne Herndon revealed the complexities of the world of biotech business. Rob Paarlberg, a friend and intellectual colleague, has written an important and insightful book (Paarlberg, 2008), from which I gleaned much about the political dynamics of biotechnology and Africa. My richest insights into agriculture in rural East Africa are due to Ruth Oniang'o (Honourable Professor Ruth Oniang'o). Ruth is a remarkable woman. For many years she was a professor of nutrition at Jomo Kenyatta University in Nairobi. She founded the *African Journal of Food, Agriculture, Nutrition and Development* and a local non-governmental organisation (NGO), the Rural Outreach Programme. She served as a member of the Kenyan parliament for one term. Working with her NGO and visiting rural areas of western Kenya have profoundly shaped my views on agriculture in Africa. The HIV/AIDs and poverty relief work of my niece, Jessica Bokhout, in South Africa and Zambia are inspiring. She read and discussed with me many of the chapters of this book. Her insights on the inner workings of NGOs are rich and nuanced. Her views on the potential harms of patents on those in low- and middle-income countries, on the attraction of organic farming and, especially, on the content in the chapter on Africa offered a helpful and needed alternative perspective. I have learned a great deal from David Castle's writings on social issues in genomics and biotechnology and from stimulating conversations over the last few years. As is always the case, this book would not have appeared without the fine work of Hilary Gaskin, Joanna Garbutt,

Anna Lowe and Christina Sarigiannidou at Cambridge University Press, and thanks to Joe Garver for meticulous copy-editing.

I owe an enormous debt of gratitude to my wife, Jennifer McShane, whom I met in high school and to whom I will have been happily married for 40 years in 2011. She has constantly supported my endeavours, endured my philosophical analysis of nearly every idea and action arising in our lives, and proofread all that I have written over the last 40 years. Although my three adult children, Eirinn, Kerry and Jonathan, and my dad, Lewis, and his wife, Pat, have not made a direct contribution to this book, their love, support and individual achievements are part of the foundation on which my own sense of self is built.

Introduction

Food and water are essential to human life; more specifically, safe water in sufficient quantities, and safe and nutritionally balanced food in sufficient quantities are essential to good health. Until the twentieth century in developed countries (rich countries), neither could be taken for granted; for most of the world's people today, neither can be taken for granted. People in rich countries, however, have for most of the last century had access to abundant, affordable and safe food and water. This is, incontestably, a direct function of advances in science and technology. Moreover, meeting the challenges of tomorrow will depend on continued advances. Jeffery D. Sachs eloquently makes this point in his book *The End of Poverty*:

> I believe that the single most important reason why prosperity spread, and why it continues to spread, is the transmission of **technologies** and the ideas underlying them. Even more important than having specific resources in the ground, such as coal, was the ability to use modern science-based ideas to organize production. The beauty of ideas is that they can be used over and over again, without ever being depleted. Economists call ideas nonrival in the sense that one person's use of an idea does not diminish the ability of others to use it as well. This is why we can envision a world in which everyone achieves prosperity. The essence of the first industrial revolution was not the coal; it was how to use the coal. Even more generally, it was about how to use a new form of energy. The lessons of coal eventually became the basis for many other energy systems as well, from hydropower, oil and gas, and nuclear power to new forms of renewable energy such as wind and solar power converted to electricity. (Sachs, 2005, pp. 41–42)

This, although completely accurate, is the rosy side. The benefits of science and technology have not been achieved without attendant problems. It is worth noting that many, but by no means all, of these problems have resulted from human inattention, greed and optimism and are not the result of advances in science and technology *per se*. Furthermore, even factoring in the problems, few people, on balance, would wish to relinquish the benefits that arise from science and technology; very few would trade the challenges of today for those of 500 years ago. Our almost universal embrace of the benefits of science and technology in medicine and dentistry – including those arising from medical biotechnology during the last several decades – provides powerful support for this view. Nonetheless, one obvious lesson from the history of science and technology is that anything less than intense and continual vigilance is irrational and imprudent. Seizing benefits and identifying and mitigating harms are inextricably connected endeavours. To believe that benefits can be seized while identifying and mitigating harms ignored is sheer folly.

Science and technology have been at the core of the success of rich countries in thwarting the prediction of Thomas Malthus (1798). Malthus claimed that human populations will, unchecked, increase geometrically while resources (food, shelter and the like) will only grow arithmetically. At some point, the population will outstrip the available resources and an intense competition for resources will ensue, leaving many with inadequate resources and, hence, desperate. For most of the twentieth century, agricultural technology advanced by employing millennia-old breeding knowledge and coupling it with contemporary population, quantitative and molecular genetics. For millennia, animal and plant agriculture relied on selecting organisms with desirable traits as a breeding stock. As new advantageous traits were identified or emerged, organisms with those traits became the new breeding stock. As scientific knowledge advanced, especially in genetics, the understanding of traits, hybridisation and selection became more sophisticated. In the latter part of the twentieth century, based on advances in cell and molecular biology, biotechnological manipulation of the genomes of organisms became possible. Governments, agencies and regulators in most rich countries approved numerous medical, environmental and agricultural applications. Of these applications, agriculture – specifically plant agriculture – became the target of intense criticism. The debate over agricultural biotechnology continues to rage and that debate is the focus of this book. Although slightly dated, the collection of articles in *Genetically Modified Foods: Debating Biotechnology* edited by Michael Ruse and David Castle (2002) provides an excellent glimpse into the differing opinions.

Engaging in the debate, obviously, involves examining scientific evidence and considerable space in this book is devoted to scientific evidence. But the things that have emerged as central in the debate are more philosophical in character. Issues, for example, about the sanctity of life and the immorality of manipulating it, the balancing of benefits and harms, the avoidance of certain kinds of harms, the ownership of new life forms, the value of biodiversity, the value of safe, affordable food and so on. Consider the claim made by Great Britain's Prince Charles in his Reith Lecture (HRH The Prince of Wales, 2000), 'I believe that if we are to achieve genuine sustainable development, we will have to rediscover, or re-acknowledge, a sense of the sacred in our dealings with the natural world, and with each other.' Lofty and eloquent as this sounds, drawing out its meaning is challenging.

What does 'genuine sustainable development' mean? Can there be ungenuine sustainable development? What is the measure of 'sustainable' and sustainable for whom or what? There are those who consider the continued loss of species as evidence of a failure to have sustainability. There are others for whom the essence of sustainable development resides in the continuation of humanity. For them, sustainable development is important – perhaps morally required – because continued human existence is under threat from a continuation of the practices of the last couple of centuries; this is a very anthropocentric motivation. There are, of course, other positions on the meaning and measure of 'sustainable' but all are philosophical in character. Furthermore, what might Prince Charles have meant by 'sacred'? Perhaps he had in mind a theological sense of the requirements of stewardship that God has given humans, and of humility that respects rather than usurps God's natural order. Or perhaps this is a thoroughly secular sense of sacred, something like recognition of the beauty and wonder of the natural world, and of the delicate balance that we can so easily disrupt. More importantly, what follows from accepting 'a sense of the sacred in our dealing with the natural world'? Surely, this is not a recommendation that we return to a way of life led by our early ancestors; caves for shelter, for example. The phrase is entirely unhelpful unless it can be given some substance. Is atomic electricity generation a violation of this 'sense of the sacred'? Is air travel a violation? Is using birth control pills a violation? Is producing recombinant insulin from bacteria a violation? In short, how will we know when we are adhering to and when violating this 'sense of the sacred'? Platitudes such as those invoked by Prince Charles are useful rhetorical devices but they do not advance rational decision-making; indeed, they frequently, as in this case, frustrate rational decision-making and lead to imprudent courses of action. This is why philosophical analysis is an essential component in any examination and analysis of socially, morally, legally and politically important issues arising from scientific advances.

To further emphasise this essential role, consider yet another example. Vandana Shiva (1997) claims:

> When organisms are treated as if they are machines, an ethical shift takes place – life is seen as having instrumental rather than intrinsic value. The manipulation of animals for industrial ends has already had major ethical, economic, and health implications. The reductionist, machine view of animals removes all ethical concern for how animals are treated to maximize production.

There is a lot packed into these three sentences. There are valuable insights and murky implications. Her main concern in this passage and in the section in which it occurs is animals – specifically agricultural animals. Beginning, however, with the phrase 'when organisms' invites one to generalise beyond agricultural animals, indeed beyond animals to bacteria, yeasts, plants and the like. In effect, she is generalising from a convincing case for agricultural animals to all organisms; her reference to 'organisms' entices the reader into accepting that her narrow claims apply to **all** organisms. I fully agree that most agricultural animals are treated appallingly and that ethical concerns are muted by a factory farm structure designed to enhance profits. Whether this is the result of a mechanistic and reductionist view is less clear but it is at least a tenable hypothesis. What does not follow is that ethical concern for 'animals' beyond agricultural animals is also removed. Cruelty to animals does occur but there is widespread public support – in rich countries at least – that such cruelty is unacceptable. Societies for the prevention of cruelty to animals abound, and research animals have for the last 25 years been protected by laws and review processes, precisely because there is little public tolerance for cruelty to animals. Without care, one can easily be seduced into accepting a view about all animals based on a narrow case for agricultural animals. Moreover, the case may seem to have been made for all 'organisms'; it has not. The importance of this latter point is that the emotive invoking of animals as machines and viewed through a reductionist lens, simply does not apply in any natural way to plants – agricultural, horticultural or other kinds – or bacteria, but they do seem to be gathered up in 'organisms' in this passage. There is a subtle analogy at work here, comparing attitudes towards, and treatment of, agricultural animals with attitudes towards, and treatment of,

all organisms. In Section 3.1 below, the value of analogy is explored, as is its abuse; Shiva's is clearly an abuse.

Furthermore, there is a significant difference between methodological reductionism (which abounds in all sciences and in medicine) and mechanistic reductionism. The latter involves accepting that **the nature of things** is such that whole entities (materials, organisms and so on) can be reduced to their parts in a way that the whole is no greater than the sum of its parts. It is not an assumption to guide research or investigation but a commitment to the ways nature is structured. I do not believe my dog is a mere machine (mechanistic reductionism) but if he is ailing, I assume, as a **method** of investigating the cause, that some part of him is not functioning properly (methodological reductionism). Shiva, as I conceded, may be correct that mechanistic reductionism is at work in the way we think about and treat agricultural animals but a biotechnologist does not have to accept this kind of reductionism (methodological reductionism is enough) to engage in genetic engineering and even if she did, it is not at all clear what the ethical implications of treating plants or bacteria this way are. By blending the two kinds of reductionism, she can slide from one to the other uncritically.

Finally on this example, there is the matter of 'instrumental rather than intrinsic value'. This is set up as a dichotomy; it is one or the other. Actually, as the discussion of Kantian ethics in Section 3.2 makes clear, it is usually both that are at work for humans as well as other animals. It is not ethically problematic to treat someone as a means (an instrument) if she is also being treated as an end (something with intrinsic value); labourers have this duality attached to them all the time. Also, the owner of a horse may well use the horse for instrumental ends – racing for prize money, for example – but also recognise that the horse has intrinsic value and needs to be properly cared for and tended: indeed, in many cases, owners confess they love their horse. Again, Shiva may be correct that pigs, poultry, cattle and such are seldom viewed by farmers as having intrinsic value but the generalisation to other contexts is again specious, as is the implication that valuing an animal instrumentally is incompatible with also valuing it intrinsically. And, how any of this applies to plants and bacteria is unclear.

Consider a final example, one that focuses on a reliable supply of food. Of late, a plethora of food movements has grown up in rich nations – nations where food is, with minor exceptions, plentiful, safe, affordable and readily accessible. The slow food movement (using fresh ingredients with dishes prepared just before serving, by contrast with fast food – e.g. McDonald's – factory prepared and prepackaged food) and the locavore movement (using ingredients grown or raised locally – e.g. the 100-mile diet) are examples. Although there are clear aesthetic, health and environmental benefits to eating locally grown food, favouring free-range animal farming, enjoying on-site preparation using fresh ingredients, and minimising prepackaged and preprocessed foods, there are also demonstrable harms, as will become apparent from the examinations undertaken in this book, especially in Chapter 7 on the organic food movement. Staying with the locavore movement, one potential harm is an inability to respond to local crop failures. A reliable, adequate supply of food requires widely distributed sources. Without this, a local population (a 100-mile-diet population, for example) risks famine from inclement weather, plant or animal disease, elevated pest populations and the like. Famine from crop failure, disease outbreaks and so on occur frequently around the world. The solution, especially in rich nations, is to import excess production from elsewhere. In a world where every community relies heavily or exclusively on local production – 'local' often extends beyond 100 miles but then so do most crop failures due to weather or pest invasions – there will be no incentive to produce food beyond local demand; modest unplanned excesses will occur from time to time but not in the quantities needed to relieve a significant famine elsewhere, and certainly such excesses cannot be relied on. So a world of local production and consumption is a precarious world, one that actually looks a lot like agriculture in low- and middle-income nations in Africa today and agriculture in Europe 300 years ago. The pattern of famine, starvation and poverty that is characteristic of African nations should make people in rich nations nervous about abandoning a global agricultural model. A healthy global agricultural marketplace is consistent with, indeed may benefit from, some level of local consumption, but eating locally cannot be the global norm without courting disaster.

Obviously, finding the right balance between local and global, price and quality, small scale and large scale is a prudent and rational approach, and is critical to successful policy and action. Finding the right balance contrasts with championing one end of a spectrum; many advocates of the 100-mile diet champion one end of the food source spectrum, thereby risking the harm outlined above. One component of the analysis undertaken in this book is the identification of end-of-spectrum views, the uncovering of their benefits and flaws, and seeking the rational balance that maximises human well-being, reliable food supply, environmental protection and sustainable agricultural practices – sustainable economically and environmentally.

These three examples draw out different facets of the same point. Philosophical analysis is an essential element of any examination of the ethical, social, legal and political aspects of issues arising from scientific advances. Failure to engage in the analysis is an abdication of reason and a ceding of the debate to mere persuasion, with confusion, an untameable cacophony of voices, and ill-considered policies, laws and attitudes. It would be disingenuous, and entirely irresponsible, not to concede, at this point, that philosophical analysis is not a panacea for these ills. The point is not that with philosophical analysis everything is rational and right but rather that without it the situation is many times worse. Philosophical analysis is one element in gaining traction on complex social issues, not the golden path to Utopia.

In the preface, I indicated that this is not an advocacy book but I obviously have positions and commitments that it would be disingenuous to deny or try to conceal. In the chapters that follow, I examine many conflicting claims, positions and arguments and the evidence given to support them. My current conclusions are favourable to agricultural biotechnology; I support agriculture shifting towards more genetic modification and it is, therefore, not surprising that the conclusions of the various examinations in the book are tilted in that direction. I also conclude that organic agriculture has a meaningful role to play. By contrast, I am quite negative on the continuation of non-GM (non-genetically modified), conventional agriculture. This is largely because of its unsustainable negative environmental impact – an impact I outline in Section 5.1. So, while this is not an advocacy book, it is also not a dispassionate, disinterested examination. I contend, however, that it is an evidence-based and reasoned examination; with issues of this importance, complexity and controversial nature, that is the most honest, helpful and rational approach possible.

To make sense of many of the touted benefits and harms of biotechnology in agriculture, a modest knowledge of the genetics underlying the technologies is helpful. For example, understanding some of the requisite conditions for, and mechanisms of, horizontal gene transfer enhances a rational assessment of the probability of such a transfer in the case of GM crops as well as the extent of harm from such a transfer – both, as made clear in Chapter 8, are essential elements of a robust risk analysis. Hence, in Chapters 1 and 2, I sketch, in as non-technical a way as possible, the core scientific underpinnings of biotechnology, and the techniques and applications found in agricultural biotechnology. In some cases, the exposition of some specific aspects of science and technology is associated with the topic for which it is most relevant. Two considerations motivate this strategy. First, Chapters 1 and 2 are designed to provide some background science and technology that is relevant to more than one topic or chapter. In addition, the intention is for those chapters to expound broad features of the science and technology rather than more specialised domains. Second, juxtaposing specific aspects of science and technology and the issue to which they are relevant permits a dynamic interaction between them. For example, the discussion of the purported harm of horizontal gene transfer benefits considerably from associating the scientific evidence with the various points raised.

The principal focus of this book is on the controversy over biotechnology in agriculture. That controversy, at this point, centres almost exclusively on plant agriculture, where most of the molecular modifications have occurred and have been commercialised. Consequently this book focuses mostly on GM plant agriculture. The controversy encompasses scientific, economic, political, regulatory, legal, ideological and theological dimensions. These are dealt with in Chapters 4, 5 and 6. A rigorous and robust examination of the various aspects of the controversy relies on analytical tools and methods. Chapter 3 describes the core tools and methods. At the heart of any analysis are reasoning and evidence; hence, I start Chapter 3 with an exposition of these. Many of the claims and arguments proffered in the controversy over agricultural biotechnology rest on ethical commitments. This is a complicated landscape. Different individuals and groups adhere to different ethical theories, and this, without care and attention to detail, will mean that they will fail to engage each other; they will be talking past each other. To use a word that has become common to describe such differences in theoretical commitments, their views will be *incommensurable* (there exists no common measure, no common assumptions). In Section 3.2, I set out the most commonly held ethical theories and note the differences among them but signal that in the context of biotechnology, there is a common measure: risk assessment. In subsequent chapters, I develop this claim of a common measure, especially in Sections 3.4 and 4.2.

Being aware of these different theories is essential to understanding many of the claims made and why those making them think they matter. It is also essential to understanding why gaining traction on an issue is so illusive.

Ultimately, I maintain, many of the issues arising from agricultural biotechnology can be examined in a way that mitigates the difficulties posed by different members and groups in a society adhering to different ethical theories. One element of this mitigation is risk analysis. Regardless of which ethical theory one adopts, many ethical, social, political and legal aspects of agricultural biotechnology require the identification of benefits and harms, an assessment of the balance of harms to benefits, and, if on balance the benefits outweigh the harms, a managing of the harms. For some ethical theories, risk assessment is fundamental; for others, fundamental ethical principles place constraints on risk analysis but do not render it ineffective or unnecessary. In Section 3.3, the various features of risk analysis are set out, including the essential role of values and goals.

One principle that some individuals and groups have elevated to a fundamental one is the precautionary principle. In its strongest version, it renders risk analysis entirely inappropriate. Few accept that strong version and, hence, few completely dismiss the relevance of risk analysis. Since the precautionary principle has been prominent in segments of the controversy over agricultural biotechnology, and because its interpretation and application interact with risk analysis, I examine it in Section 3.4.

Many who reject molecular biotechnology in agriculture look to organic agriculture as the alternative. In Chapter 7, I look in some detail at this alternative and the claims made about it. The thrust of the chapter is that organic is best contrasted with conventional agriculture and that the contrast with GM agriculture is unhelpful and contrived. If we are to escape the environmental ravages of conventional agriculture, GM and organic agriculture will have to be embraced. To put the view I support in its strongest terms, the antipathy towards GM agriculture expressed by those who support organic agriculture is irrational; conventional agriculture should be the target of their antipathy.

The low- and middle-income countries, in various ways at different times, have suffered at the hands of developed (rich) nations. The impact of rich countries' squabbling over GM agriculture is but another instance. Some low- and middle-income countries are slowly breaking the continuing colonial hold of rich nations, a hold that no longer depends on military subjugation but on economic control through vehicles such as trade. Sadly, that hold is also maintained by the views and actions of NGOs on whom poor nations and their impoverished citizens depend for assistance. This is sad because most of us financially support those NGOs, volunteer our time, or accept employment with them because bettering the lives of the poor matters to us. The low- and middle-income countries about which I know the most and on which the impact of rich nations' squabbles have had the greatest negative impact are in Africa. It is a vast continent and its nations differ substantially in their resources, needs and abilities. Despite billions of dollars in aid and the activity of countless NGOs, the data on poverty and health are appalling and progress is illusive. In Chapter 8, I examine the promise of agricultural biotechnology for African nations and indicate the negative impact the debate over it in rich countries has had on poor Africans. I also highlight, again, in this context the hypocrisy of rich countries around biotechnology in agriculture, medicine and environmental amelioration.

1 Scientific background

1.1 Population genetics

Although the current debate about agricultural biotechnology is often narrowly focused on molecular biotechnology (molecular genetic modification), the technological application of biology in agriculture predates the advent of molecular biology. For more than 10,000 years humans have been manipulating the traits of animals and plants (Mazoyer and Roundart, 2006; Thompson, 2009) by manipulating their genes and, thereby their genomes (the specific combination of genes in an organism's cells); the dog was likely the earliest animal to be domesticated (about 16,000 years ago). Early domestication of agricultural animals and plants was based entirely on crude experimentation (trial and error). Biological knowledge was elementary; humans learned early that offspring resemble parents, that selecting animals and plants with desirable traits and breeding them created a population of animals with those traits, and that occasionally a new trait seemed to appear. Although elementary, and based entirely on experience, this knowledge was sufficient to allow the domestication of numerous plants and animals. A biological understanding of the observed phenomena did not exist until the middle of the nineteenth century; that is, until the development of a theory of genetics. The area of genetics developed first was population genetics. Beginning in the early part of the twentieth century, it, along with quantitative genetics,[1] which will

[1] Even though I deal with population genetics and quantitative genetics in separate sections, they are closely related. Both focus on trait variation in phenotypes and both trace their origins to J. B. S. Haldane, Ronald A. Fisher and Sewall Wright. They differ mostly in the kinds of traits on which they focus. Population genetics, for the most part, concentrates on single locus traits; quantitative genetics concentrates on traits involving multiple loci and multiple environmental factors. To some extent, population genetics could be subsumed under quantitative genetics as a limiting case.

be discussed in the next section, made possible important and far-reaching modifications of plants and animals.

Population genetics and quantitative genetics are important in their own right in agriculture since the technological application of biological knowledge in these domains continues to be used extensively in plant and animal agriculture. Selecting agriculturally useful traits of plants and animals and developing populations with those traits through breeding involves, principally, the application of population and quantitative genetic theory. Furthermore, many agriculturally desirable plants are hybrids (created by interfertilising plants with different genetic profiles). Understanding the population and quantitative genetic basis of modern agricultural hybridisation is essential to advances in hybridisation. Both conventional trait selection and hybridisation continue to occupy a significant market share. Indeed, in plant agriculture, where the proportion of genetically modified (GM) seeds planted has seen a steady increase, it is still the case that hybrid and conventional seeds are supplied and planted in abundance; data collected and analysed by Precision Agricultural Services, Inc. and reported by Monsanto (2010) indicated that in 2010 for corn seed alone there were more than 6,000 traited hybrids and over 1,000 conventional seeds offered for planting. Of special importance to organic farmers, population genetics and quantitative genetics are also essential to understanding the characteristics of 'open pollinated' plants, which make collecting and retaining seed from year to year feasible. Hence, even with the advent of molecular genetic modification, population genetics and quantitative genetics continue to be important. Moreover, they are important to aspects of GM seed production and GM agricultural practices. For example, a technique for inhibiting the development of insect resistance to a pesticide expressed by some GM plants relies heavily on population dynamics (the combining of population genetics and ecology), a technique which I describe in more detail in Chapter 6.

The development of contemporary population genetics began with a brilliant and seminal, but at the time largely unnoticed, contribution by Gregor Mendel in 1865. Mendel was interested in hybridisation in plants (interfertilising two varieties of a plant) and set out to discover what happens in subsequent generations of intrabred hybrids. His explicit goal was to discover generally applicable laws. Although knowledge of hybridisation predates Mendel, it was not until his work that the underlying mechanisms were discovered. In the earliest period of agriculture (the Neolithic period approximately 10,000 years before the present), the goal was to avoid hybridisation (Mazoyer and Roundart, 2006). Today, some of the most beneficial traits, including yield improvement, result from controlled hybridisation based on robust biological knowledge.

Mendel's work attracted little attention until the beginning of the twentieth century. In what is now seen as an ironic twist of fate, Darwin's theory of evolution, as set out in 1859 in *On the Origin of Species*, assumed the hereditary transmission of traits but he had no credible theory of heredity; he relied instead on the wide acceptance of observed trait inheritance. Had Darwin, or any of his colleagues for that matter, known about Mendel's theory, he could by the fourth edition (1866) have included it and further strengthened his case. Early work on Mendel's theoretical model concentrated on its implications and on extending the scope of the model. Mendel provided a mathematical model that described a causal mechanism which accounted for the phenomena he observed. Advances in the optics of microscopes and in staining techniques made possible, during the period 1840–1900, increasingly clearer observations of the behaviour of what today we call chromosomes. In 1902, Walter Sutton, a postgraduate student at Columbia University, in a single offhand sentence, connected the observed behaviour of chromosomes with Mendel's mathematical account of his hereditary factors.

> I may finally call attention to the probability that the association of paternal and maternal chromosomes in pairs and their subsequent separation during the reducing division as indicated above may constitute the physical basis of the Mendelian law of heredity. To this subject I hope soon to return in another place. (Sutton, 1902, p. 39)

Subsequently, in 1903, he provided a more detailed account (Sutton, 1903; see also Crow and Crow, 2002). Although this was a controversial hypothesis in 1902, by 1910, the hypothesis had received considerable experimental and theoretical support.

The next major contribution to population genetics was made independently by G. H. Hardy (Hardy, 1908) and Wilhelm Weinberg (Weinberg, 1908). Both provided a formulation of an equilibrium state for a Mendelian population (i.e. a population that conforms to Mendel's model). In essence, the formulation states that the ratio of Mendel's factors (today called alleles) will remain constant in all subsequent generations after the first unless something like selection, mutation, immigration, emigration and the like occurs; so unless something happens, the allelic ratios will remain constant forever. Of course, in actual populations, the ratios do change from generation to generation, entailing that one or more of selection, mutation, immigration, emigration and the like are occurring. Subsequently, this equilibrium principle was incorporated into contemporary population genetics, which coalesced in the 1920s with the work of J. B. S. Haldane (Haldane, 1924–32, 1932), Ronald A. Fisher (Fisher, 1930) and Sewall Wright (Wright, 1931).

The nuclei of cells contain chromosomes (cells with a nucleus are called eukaryotic; those without, prokaryotic). Chromosomes exist in matched pairs when a cell is not undergoing division, a phase known as the resting phase. Cells engage in two kinds of division: mitosis and meiosis. Mitosis results in two cells each identical to the parent cell; each has a complete set of the original matched pairs of chromosomes. Meiosis results in four cells, the nuclei of which have only one set of the original matched pair of chromosomes. These cells are called gametes; human sperm and ova are gametes. During the process of fertilisation gametes from males and females combine to create a new single cell, the nucleus of which has a complete set of matched pairs of chromosomes; normally this cell undergoes mitotic division numerous times, resulting in a mass of identical cells. At this point, these cells are stem cells; stem cells are generic cells and have the property of being able to transform into any of the specific cells of the adult organism (e.g. heart, liver and skin cells). Once transformed, further mitotic division produces only the specific type of cell it has become. This is why stem cells are so valuable for current medical research and why embryos in the early stages of development are an important source.

Particular locations on chromosomes give rise to different traits (characteristics) of the adult organism (its phenotype). The processes through which those traits arise during embryological development are complex and still not completely understood but it is now clear that the basic genetic code for the organism is embodied in that organism's chromosomes. What is unclear is how that code gives rise to the adult organism. Much is known but the process is complex, involving some genes controlling the expression of others, environmental conditions, sequencing and many other aspects; there is still much to be discovered. A point of terminology – I hereafter will use the term 'development' to cover the process through which an adult organism arises. Hence, it covers the period from fertilisation up to the adult plant or animal.[2]

Some characteristics (aspects, traits) arise from the genetic code found at one location on one chromosome (sickle-cell anaemia, for example); most, however, involve many locations on many chromosomes and are influenced by many factors during development. The more closely a trait can be tied to one, or a very few, positions on a chromosome, the more straightforward and efficacious is the genetic manipulation required to alter, remove or introduce that trait.

Let's look a little more closely at Mendel's postulation of hereditary 'factors', which in contemporary population genetics are called alleles. Two alleles are associated at each location (locus) on a matched pair of chromosomes; a matched pair of alleles is a gene. The number of possible combinations depends, of course, on the size of the set of alternate alleles. If only one kind of allele can occupy that location, then every organism will have the same pair of alleles (say, *AA*) and each member of the pair will be identical. If two alleles can occupy that locus, there will be three possible unique pairings (*AB*, *AA*, *BB*); *AB* and *BA* are not unique combinations and constitute identical genes. If three alleles can occupy the locus, there will be six unique combinations (*AA*, *AB*, *AC*, *BB*, *BC*, *CC*). As the number of possible alleles at a locus increases, the number of genes increases. As the number of possible genes at a locus increases, the number of traits by which the adult organisms can differ from each other increases.

At any point in time, the proportion of a given allele in the population can be determined. In a simple case with two alleles *A* and *B* at a locus, *A* may be more numerous than *B* (for example, the ratio of $A{:}B = 7{:}1$). For mathematical convenience, the proportions are normalised to sum to 1. So the ratio 7:1 is normalised to 7/8:1/8 or 0.725:0.125. An example of an allelic pairing that yields that ratio is:

$$20\ AA{:}1\ AB{:}1\ BB$$

[2] There is, obviously, no precise point at which an organism is an adult. From an evolutionary point of view, ability to participate in the production of offspring marks adulthood. From a social point of view, as in the case of humans, it occurs somewhat later, ranging from 18 to 25 years of age.

The *AA* combination contributes 20 *A*s; the *AB* combination contributes 1 *A*, for a total of 21 *A*s. The *AB* also contributes 1 *B*, which along with the 2 *B*s contributed by the *BB* combination results in 3 *B*s. Hence there are 21 *A*s and 3 *B*s. Dividing both by 3 yields 7 *A*s to 1 *B* ($A{:}B = 7{:}1 = 0.725{:}0.125$). What G. H. Hardy and Wilhelm Weinberg demonstrated was that in every generation after the first, the proportion of alleles at a locus, in a closed population, will be the same – an equilibrium will be reached. That equilibrium can be disturbed in open populations – populations open to selection, immigration into and emigration from the population, by meiotic drive (where gametes are not produced in equal quantities: e.g. more gametes with XX chromosomes (female) than XY chromosomes (male) are produced during meiosis) and so on. What the Hardy–Weinberg equilibrium states is that if nothing happens, nothing happens. This might seem trite (perhaps even ridiculous) but, in fact, it is a powerful principle. Since they proved that if nothing, except random mating, is occurring in the population, the allelic ratios will remain constant over time, if there is a change in the ratios, something must be happening to cause the change; there must be an explanation in terms of some factor(s) perturbing the system.

The proof of the Hardy–Weinberg equilibrium is straightforward. Assume a locus with two alleles *A* and *B*; also assume, in the founding generation F_0, $p =$ the proportion of *A* alleles and $q =$ the proportion of *B* alleles. Construct a breeding matrix (assuming random mating) as follows:

	$p(A)$	$q(B)$
$p(A)$	$p^2(AA)$	$pq(AB)$
$q(B)$	$pq(AB)$	$q^2(BB)$

AB is the same as *BA*, so there will be $2 \times pq$ of this combination. Hence, the ratios after mating (i.e. in the next generation, F_1–F_n designates the *n*th generation with F_0 being the founding generation) are: $p^2AA{:}2pqAB{:}q^2BB$. So, summing the *A*s and *B*s yields, $A = 2p^2 + 2pq$ and $B = 2q^2 + 2p$; hence, $A{:}B = 2p^2 + 2pq{:}2q^2 + 2pq$. Dividing both sides of the right-hand ratio (i.e. the *p* and *q* side) by 2 yields $A{:}B = p^2 + pq{:}q^2 + pq$. Factor each side of the ratio to yield $p(p + q)A{:}q(q + p)B$. Normalise this ratio, so that, $p + q = 1$ (hence, $p = 1 - q$ and $q = 1 - p$), by replacing *q* on the left side with $1 - p$ and *p* on the right side with $1 - p$, which results in the ratio $p(p + (1 - p))A{:}q(q + (1 - q))B$ or, removing the unnecessary parentheses, $p(p + 1 - p)A{:}q(q + 1 - q)B$. The *ps* in the parentheses on the left cancel, leaving $p(1)A$, and the *qs* in the parentheses on the right cancel, leaving $q(1)B$; since multiplying by 1 changes nothing, the F_1 generation ratio is, $p(A){:}q(B)$. This was the starting ratio in the F_0; hence, the ratio after mating remains unchanged.

The Hardy–Weinberg equilibrium plays a role in population genetics similar to the role played by Newton's first law in Newtonian mechanics. Newton's first law states that all bodies remain in constant rectilinear (straight line) motion or at rest unless acted upon by an external, unbalanced force. That is, if nothing happens, nothing will happen; the state of the system will remain the same forever. Hence, if an object undergoes negative or positive acceleration, or takes any path other than a straight line, a force must be acting on it. If the allelic ratios in a population change, something must be acting in or on that population.

In addition to postulating factors (alleles), Mendel, to explain fully his experimental results, had to postulate a property of his factors: factors could be dominant or recessive. Here's how this property is put to work in the theory. As indicated, Mendel's experiments were designed to explore hybridisation. Beginning with seeds that bred true for a trait (Mendel explored seven pairs of traits[3] but the one most often used in explications of his work is wrinkled and round peas), Mendel cross-fertilised the true breeding plants (e.g. ones that always yielded round peas and ones that always yielded wrinkled peas) to produce hybrid plants – pollen from round peas was used to fertilise ovules from wrinkled peas and vice versa. What he found was that in the first generation all the plants had the same trait (e.g. always produced round peas). When he crossed the offspring of this first generation, he found that some plants manifested one trait, and others the other trait (e.g. some produced round peas and others produced wrinkled peas); the ratio was 3:1 (e.g. 3 round to 1 wrinkled).

[3]
1. Round vs. wrinkled peas
2. Yellow vs. orange peas (seen through transparent seed coats)
3. Seed coats white vs. grey, grey-brown, leather brown
4. Smooth or wrinkled ripe seed pods
5. Green vs. yellow unripe seed pods
6. Axial or terminal flowers
7. Long vs. short stem (he chose 6–7 ft and $\frac{3}{4}$–$1\frac{1}{2}$ ft).

To explain these results, he postulated that his factors (one responsible for round peas, another for wrinkled peas) segregated when gametes are produced – just as chromosomes were later discovered to segregate during meiosis. If all the factors are the same in all the breeding plants, all the gametes will have the same single factor (*S* – smooth – for example). When the two gametes are united, the zygote will have two identical factors for that trait (e.g. *SS*); these organisms are called homozygous or homozygotes. Those plants will breed true generation after generation. Hybrids, however, will have one factor from plants breeding true for a trait *S* and one from plants breeding true for a different trait *W* (wrinkled). The hybrid zygote will be *SW*; these organisms are called heterozygous or heterozygotes. When *SS* plants are crossed with *WW* plants, all the offspring will be *SW*. So why, in the first generation (designated F_1, the original generation being F_0), did all the plants manifest only one of the traits when they all had an allele for each trait? Because, postulated Mendel, *S* dominates over *W*, so when they are together in a combination the trait *S* will always dominate and be manifest in the plant. The next thing to be explained is why, when the hybrids of the F_1 generation were interbred (creating generation F_2), were both traits found, and found in the ratio 3:1. The explanation is mathematically simple. When two hybrids are bred, some zygotes will be homozygous for each of the factors and others heterozygous. Since all the plants in F_1 are *SW*, each will produce, on average, 50 per cent *S* and 50 per cent *W* gametes. Using an elementary matrix product, the 3:1 ratio is obvious.

		Gametes of plant A	
		S	*W*
Gametes of plant B	*S*	*SS*	*SW*
	W	*WS*	*WW*

The combinations (e.g. *SS*) are the product of combining the relevant gametes from plant A with relevant gametes from plant B.

The same thing can be illustrated diagrammatically.

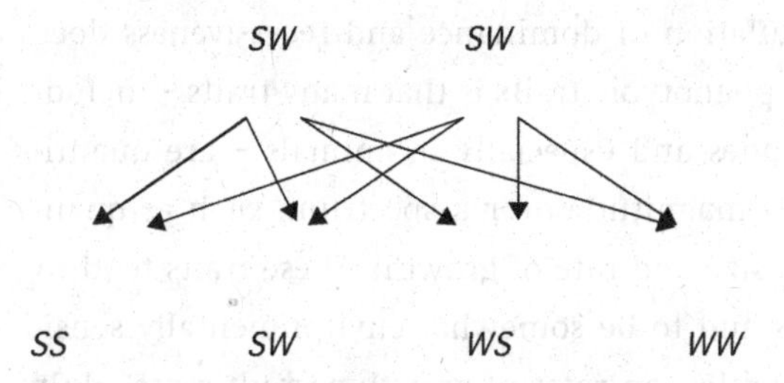

As the matrix and the diagram demonstrate, the possible re-pairing of gametes from two hybrids are *SS*, 2*SW* (*SW* + *WS*), and *WW*. Since *S* is dominant, the 2*SW* will manifest the *S* trait as will the *SS* because it is homozygous for *S*. Only *WW* will manifest the *W* trait. Hence three of the four combinations will manifest the *S* trait and one will manifest the *W* trait (i.e. *S*:*W* = 3:1).

Although Mendel's postulation of dominant and recessive factors (alleles) is conceptually important, it does not provide a complete basis for understanding phenotypic traits. Frequently, heterozygotes do not manifest one of the discrete traits found in the contributing homozygotes. For example, a phenomenon called heterozygote superiority[4] occurs when a phenotypic property of the heterozygote makes it fitter than either homozygote – as in the case of a person with an allele for sickle-cell haemoglobin and an allele for normal haemoglobin. The homozygote for normal haemoglobin is susceptible to malaria and the homozygote for sickle-cell haemoglobin is susceptible to sickle-cell anaemia; the heterozygote is resistant to malaria and does not develop sickle-cell anaemia. Fitness is always relative to an environment – the sickle-cell heterozygote is fitter in an environment where malaria is endemic, for example. In agriculture, the environment is, in large part, created by humans, and agricultural crops and animals are fit relative to that environment (an environment determined by the needs and interests of farmers, food processors, shippers, consumers and so on). Many agricultural crops (e.g. wheat, rice, corn/maize) are the product of human manipulation of reproduction to create novel hybids because the traits of these hybrids are superior to those of either homozygote (more on this in Section 1.3).

[4] Heterozygote inferiority also occurs (Christiansen, 1978).

1.2 Quantitative genetics

Another reason Mendel's postulation of dominance and recessiveness does not fully account for observed phenotypic traits is that many traits – including agriculturally significant ones and especially in animals – are quantitative traits (traits that vary in magnitude over a spectrum, such as quantity of milk production, udder size and rate of growth). These traits tend to be the product of many genes and to be somewhat environmentally sensitive (such as the impact of nutrition on rates of growth and ultimate adult height). Quantitative traits vary by degree over a spectrum because of the multiple genes involved in the development of the trait. In cases where a trait is controlled by a single locus, a single allelic substitution can produce a large difference in the trait. When multiple genes are involved, a single allelic substitution will produce smaller differences, leading to a gradation in magnitude.

An important property of many quantitative traits is the effect of the interaction of the genes that control the trait; these are known as epistatic effects. In simple cases, a trait can be the product of many genes without any interaction among the genes other than the additive effect they each contribute to the trait. When, however, genes interact (such as one suppressing the expression of another), the magnitude of the trait will depend not only on the contribution of the particular allelic combination at each of the relevant loci but also on the particular mix of these allelic combinations. Abstractly, this can be illustrated by considering two loci, each of which has two alternate alleles (*A* and *a*, *B* and *b*). If no epistasis occurs, the differences in organisms will be the additive effect of the *different* combinations of the alleles at each locus. If epistasis occurs, *AaBb* and *Aabb* could be different not just because *Bb* has a different effect on the trait than *bb* but also because *bb* has a different effect on *Aa* than *Bb* does. *Bb*, for example, might inhibit the effect *Aa* can have on the trait, whereas *bb* allows the full expression of *Aa* on the trait. In more complex cases, say, four loci A, B, C, D, a particular allelic combination at B (say, *bb*) might inhibit the expression of gene A but a particular allelic combination at D might inhibit the effect of *bb* on A. Epistasis clearly broadens dramatically the possible effects of genes on a trait; add to this the fact that many loci have more that two alternate alleles and it is easy to see how a trait could manifest a large array of magnitudes that create a continuous or quasi-continuous spectrum for that trait.

The spectrum is quasi-continuous when trait variation is discrete but, in a population with a large number of potential phenotypes, it is effectively continuous. Consider the number of hairs on a dog. Hairs can be counted and, hence, there is a discrete numerical value in increments of 1. However, if the potential number of variants is large, say, 10,000, then the scale appears continuous. The essential feature of quantitative traits is that they are the product of multiple genes and are sensitive to environmental factors; whether the scale for the trait is discrete or continuous depends on the trait. Three types of quantitative traits are often identified: threshold traits (the trait is either present or not and is hence discrete), metric traits (the trait variation is continuous – all values on a continuous scale can, in principle, be realised), and meristic traits (the trait measurement is a discrete quantity but a large number of discrete variants are possible). Weight, height, total skin area and the like are examples of metric traits. The number of body hairs and the number of ova in the ovaries just prior to the onset of menses are examples of meristic traits. Being left-handed and having a cleft palate are examples of threshold traits. In an agricultural context, the volume of milk produced is a metric trait. The quantity of wool, on the other hand, depends on the number of follicles, which is discrete with a very large number of possible values; it is a meristic trait.

Separating the genetic determinants from the environmental ones is challenging. One manifestation of the brilliance of Ronald A. Fisher, who, you will recall, was a founder of modern population genetics, was his experimental method (see Fisher, 1935). Much of Fisher's research was in agriculture; his experimental method was founded on three elements: randomisation, replication and blocking. Essentially, the method requires the experimenter to divide a field into paired adjacent blocks and to manipulate the environmental variable (adding nitrogen fertiliser, for example) in one block but not the other. The block to be manipulated is chosen through a random process. Since there will be many such paired blocks in the field, replication is achieved. Because the blocks are adjacent, it is reasonable to assume that they are homogeneous in all respects except the experimental variable. Any differences found (statistically significant differences) can be attributable only to the experimental variable and, hence, it can be declared the cause. Although this method is commonly used in agriculture, the most commonly encountered references to this method today are not in agriculture but in medicine, where it has been touted as the gold standard of evidence. This is unfortunate because Fisher's experimental method is ideally suited to agriculture but not to clinical trials in medicine. In clinical trials, the method is known as randomised, controlled trials (RCTs). The critiques of RCTs in medicine are legion and I have set out the major ones in several publications (Thompson, 2010a, 2010b).

That many of the traits of animals are quantitative makes the process of trait selection complicated. Compounding this complexity is the fact that in most cases more than one trait is desired; this is also true of agricultural plants. Charles Smith (1998) has identified 30–40 traits in dairy cattle, for example.

1.3 Hybridisation

Open pollinated plants are those that will breed true from generation to generation. They may have been manipulated, through selection or even molecularly, to fix certain beneficial traits; the criterion for open pollination is simply that the plant breeds true. This is an important feature for those who wish to retain seed from one season to the next, a point to which I return later. Hybrids, by contrast, will not breed true in the next generation. Consider the simple case of a plant heterozygous at a locus; here I focus on plants but the same things are applicable to animals as well. During meiosis (gamete formation), pollen and ovules with only one of *A* or *a* will be formed. The ratio of *A* pollen and *A* ovules to *a* pollen and *a* ovules is close to 0.5*A*:0.5*a*. Assuming close to random pollination, the segregated *A* and *a* alleles will recombine in the fertilised ovules in this way:

	A	*a*
A	*AA*	*Aa*
a	*Aa*	*aa*

Hence, a field of hybrids will produce 50 per cent non-hybrid seed (the *AA* and *aa* combinations). A farmer will not know by inspection which are the hybrid seeds. Only by germinating the seed and growing the plants can one tell, and were a laboratory procedure available, it would have to examine each of the seeds to sort them into *AA*, *Aa* and *aa* – a procedure that would be complicated, expensive and time-consuming. Hence, a farmer who wants to grow a plant that is heterozygous at that locus will, each year, need to buy the seed from a seed company. Seed companies guarantee that close to 100 per cent of the seed will be heterozygous at that locus because they maintain and cross-fertilise original homozygous plants.

This is, of course, a simple example in which there is only one heterozygous locus but it illustrates the more general feature of hybrids. The genetics in actual cases is far more complex than a single-locus model; additivity, dominance and epistasis (effects between loci) are all important. Also, frequently, desired traits are quantitative (involving more than one locus and environmental factor) and commercial hybrid seed often involves creating hybrids from varieties found in different populations and the desired trait is only found in the hybrid. An in-depth account of the quantitative genetics of line crosses is provided by Lynch and Walsh (1998). Agriculturally beneficial hybrids are frequently obtained by crossing separate varieties, varieties which would not naturally interfertilise. Several outcomes are possible when creating hybrids by crossing plants from different populations; the seed may fail to develop, it may develop but produce a malformed plant, it may produce a normal plant that lacks vigour, it may produce a vigorous mature plant that is sterile, or it may produce a viable mature plant that will reproduce. For agricultural purposes, it is the viability and vigour of the plant and its agriculturally desirable traits that are important. Hence, sterility is only an issue if a farmer wants to retain seeds. This is unlikely, because, like the single-locus example, the offspring will be a mix of hybrids and non-hybrids.

Hybrids are agriculturally valuable because they can manifest a trait not found in either parent or manifest an enhancement of a trait over its parental expression. One important trait found in many hybrids is greater vigour than either parent – a phenomenon known as hybrid vigour or heterosis. Hybrid maize (corn), for example, exhibits heterosis. The genetics of heterosis is still being uncovered but the phenomenon has been known for a long time; Darwin discussed it in his *The Effects of Cross and Self Fertilisation in the Vegetable Kingdom* (Darwin, 1876). What has also been known for a long time is that F_1 generation heterosis is mostly lost in the F_2 generation and beyond (remember that F_0 is the parental generation, F_1 the hybrid resulting from the cross, and F_2 the generation resulting from the reproduction of the F_1 generation), and in some cases the F_2 plants are less fit that either F_0 parent. Hence, the only way to ensure that plants will exhibit heterosis in each field planting is to use only seed produced by crossing F_0 parents. Again, seed companies maintain and cross the original parent stock to produce seeds guaranteed to be F_1 hybrids with the desired heterosis.

Maize[5] is a superb example of the agricultural benefits derived from hybridisation. In addition, it is an important agricultural crop in much of the world; many rich and middle- and low-income countries have come to depend on maize for human consumption (as kernels, starch, oil and sugar) and animal fodder. Hence, understanding the features of this crop pays many dividends. Maize is a New World crop although there are Old World relatives of maize and perhaps in the very distant past the ancestors of New World maize (*Zea mays*) were more closely related to Old World Maydeae, but, as Mangelsdorf (1974) has noted, 'The fact that corn can be crossed with both of its New World relatives, teosinte and *Tripsacum*, shows that the three taxa are related. The fact that it has never been successfully crossed with any of the Old World Maydeae strongly suggests that its relationship to them is more remote.' Contemporary maize is, hence, certainly of New World origin. In the late fifteenth century, when Europeans arrived in the Americas, it was being grown as a food crop throughout the Americas. Maize was a staple food throughout a large geographic area of South America well before Europeans arrived. Moreover, in the complete absence of a knowledge of nutritional components of food, civilisations and groups that relied heavily on maize had figured out that obtaining a complete complement of nutrients depended on combining maize with other plant-derived foods; in most cases in South America beans and squash were the complementary foods. As we know today, maize is deficient in the amino acids (see below) tryptophan and lysine and the vitamins riboflavin and nicotinic acid. Beans contain adequate quantities of all of these. Maize is also low in fat and vitamin A. Squash provides the required additional amounts of both (Mangelsdorf, 1974, pp. 1–2; McGee, 1997, p. 242). There are five types of corn grown today:

> There are five different kinds of corn, each characterized by a different endosperm composition. Pop and flint corn have a relatively high protein content and a hard rather waxy starch. Dent corn, the variety most commonly grown for animal feed, has a localized deposit of soft waxy starch at the crown of the kernel, which produces a depression, or dent, in the dried kernel. Flour corn, with little protein and mostly waxy starch, is grown only by Native Americans for their own use. What we call Indian corn today are flour and

[5] 'Corn' is a term used exclusively to denote maize in the USA. It has a broader meaning in Europe and in other English-speaking countries, sometimes being used as an alternative to 'kernel', or to 'grain' (as in 'corning' – curing with grains of salt). Sometimes, too, as in Great Britain, it designates the dominant local grain.

> flint varieties with variegated kernels. Finally sweet corn, very popular as a vegetable when immature, stores more sugar than starch, and therefore has translucent kernels and loose, wrinkled skins (starch grains refract light and plump out the kernels in the other types). It appears that popcorn was the first kind of corn to be cultivated, but all five were known to Native Americans long before the advent of the Europeans. (McGee, 1997, p. 241)

Carl Linnaeus (also known as Carl von Linné), the father of modern taxonomy, gave it the binomial name *Zea mays* (binomial = two-name structure, a genus name, *Zea*, and a species name, *mays*).

The goal of maize breeding, as with all agricultural breeding, is to maximise desirable traits: nutrients, yields, storage, days to maturity and ease of harvesting, for example. Simultaneously maximising all the valued traits is hardly ever possible; increasing the nutritional profile of a plant could entail forgoing longer storage, for instance. Selecting plants that manifest the maximum value for a trait of interest (yield is always agriculturally important) and using them as the breeding stock is an ancient and effective technique for maximising a trait. The limit of this technique is the existing maximum value. Open pollinated plants have throughout agricultural history been improved (improved relative to human goals) by this technique. Another technique is hybridisation. Its advantage over selection alone is the development of new traits or new maximum values for existing traits.

The beneficial traits are different for different types of maize. Obviously, traits affecting the popping process and product are central to popcorn and traits affecting sweetness are central to sweet maize. Yield, as already indicated, is important to all types of maize since it is a fundamental economic factor. Within each type of maize, there are numerous varieties. Crossing these varieties has proved to be an extremely successful way to improve a number of the desirable traits in maize. One trait directly related to yield is vigour (strong, healthy growth). Vigour means the plant is less susceptible to environmental stress, disease and pests; yields are consequently higher. Hybrid maize almost always manifests heterosis (hybrid vigour). Yield (kilograms/hectare, kg/ha, or bushels/acre, bu/ac) is a ready-made metric for quantifying vigour. There is a wealth of data on heterosis in maize using yield as the metric. Research conducted in the Corn Belt of the USA demonstrated dramatic yield increases from crosses of maize adapted to the Corn Belt climate with those from South America. The mean yield of the hybrids was, on average, 71 per cent higher

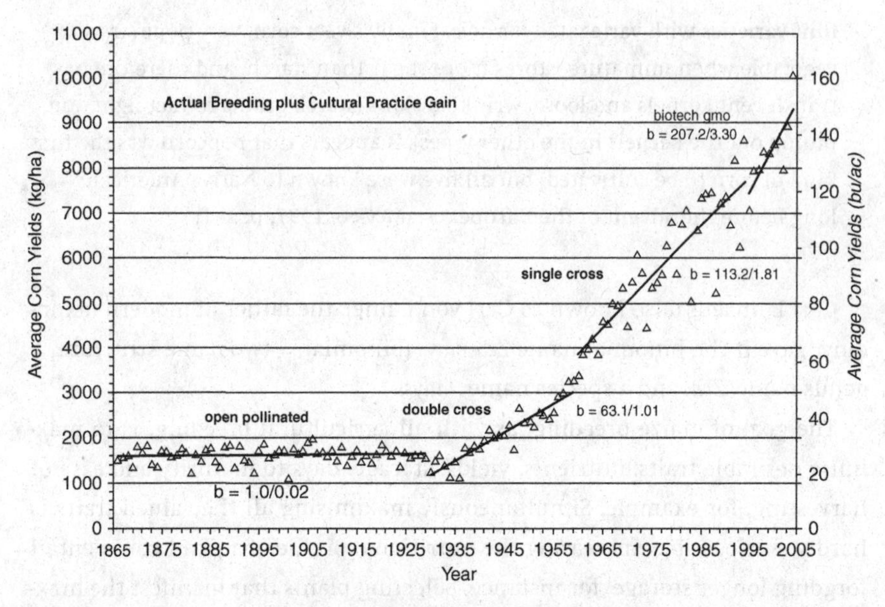

Figure 1.1 Average US corn yields and kinds of corn (from Troyer, 2006 based on data from USDA/NASS: see USDA/NASS, 2009). Reproduced with permission of *Crop Science*. *b* values (regressions kg bu^{-1}) indicate production gain per unit area per year; biotech gmo designates molecular-biotechnology-generated, genetically modified organisms (plants).

than the mean yields of the parents. For example, the Saskatchewan variety is the highest yielding in that region at 3,120 kg/ha compared to a yield of 5,310 kg/ha for the cross of Syzldecka and Motto varieties (see Hallauer, 1978, p. 233).

Figure 1.1 plots the increases in maize yield from 1866 to 2005 in the USA. The gains plotted are breeding plus 'cultural practices'. As can be seen, open-pollination varieties, even with selective breeding improvements, resulted in very low yields relative to contemporary yields from hybrid crosses (and more recently biotech). Also, open-pollination varieties had reached a plateau by 1866; the improvements possible by selective breeding alone had been wrung out of the system. It is important to be clear that many other things contributed to the dramatic yield increases from 1930 onwards – the 'cultural practices' component of the gains. Synthetic fertilisers became available, as did herbicides and pesticides. Nonetheless, factoring these out, hybridisation dramatically improved yields. It is also worth noting that maize production in the USA increased from 2 billion bushels in the early 1930s to

11.8 billion in 2006 **while the land area planted in maize *decreased* by 22 per cent.**

An important point that has been emphasised here, and which I shall continue to underscore, is that farmers who wish to obtain a benefit from hybrid plants will need to buy seed from a seed company every year. As conceded, a farmer could do what seed companies do. She could maintain sufficient stock of the parents, keep them in isolation (to avoid accidental cross-fertilisation with the hybrid crop), ensure that intrafertilisation cannot occur, ensure adequate cross-fertilisation (manually or via a pollinator such as a bee), and reserve some portion of the parental stock for intrafertilisation for the next cycle of the hybridisation process. The reality is that for most farmers this is not a cost-effective use of time or resources, not to mention that developing the skill and knowledge required is not a trivial investment. Furthermore, seed companies invest significant amounts in research and development to continually enhance their products, making it even more advantageous for a farmer to buy seeds annually.

1.4 Molecular genetics

The birth of molecular genetics dates from 1953 when James D. Watson and Francis H. C. Crick sent a letter to *Nature* setting out their conception of the molecular structure of deoxyribose nucleic acid (DNA: now more frequently cited as deoxyribonucleic acid) (Watson and Crick, 1953a). A longer article by Watson and Crick exploring the implications of the structure of DNA was published in *Nature* the following month (Watson and Crick, 1953b). Since Watson and Crick submitted the letter and paper to *Nature*, they are credited with the actual discovery. However, the 1962 Nobel Prize in physiology or medicine was awarded to Watson, Crick and Maurice Wilkins. Wilkins was awarded one-third of the prize because of the role his X-ray diffraction studies played in the discovery. Rosalind Franklin, whose X-ray diffraction studies, it is often claimed, were more directly used by Watson and Crick, had died in 1958. Since only living persons can be nominated for the Nobel Prize, she was not among the nominees.

Many researchers were on the quest for a model of the structure of DNA; Linus Pauling, already a Nobel laureate for his discovery of the alpha-helical structure of proteins (Pauling *et al.*, 1951), started with a triple helix model but was zeroing in on a model identical to that of Watson and Crick. Watson published in 1968 a delightfully frank personal perspective on the race to discover DNA's structure; it was published by Atheneum (and simultaneously by McClelland and Stewart Ltd in Canada) after the Harvard Corporation rejected it, overruling the university's Board of Syndics, which had already accepted it (Sullivan, 1968).

The chemical structure uncovered by Watson and Crick – using crystallographic data (X-ray diffraction patterns of crystals) from the work of Rosalind Franklin and Maurice Wilkins – is reasonably simple but its biological implications are deep and far-reaching. Metaphorically, DNA is like a twisted ladder. The chemicals comprising the rungs are called nucleotides; there are four of them: adenine (A), cytosine (C), guanine (G) and thymine (T). Each rung is composed of two of these nucleotides. The rungs are joined together by a polymer (a chain of repeating chemical units called monomers). This creates the sides of the ladder (the strands). The specific polymer of DNA is a sugar phosphodiester polymer. The rungs constitute a code; actually, there are two codes: a code for DNA replication and a code for protein construction.

The first code (DNA replication) depends on a chemical property of nucleotides: A can only combine with T and vice versa, and C can only combine with G and vice versa. Hence, if this metaphorical ladder is split down the middle, one half allows the construction of the other half. That is, if the nucleotide sequence on the rungs of one half is AAGTCG, since AT and CG are the only chemically possible combinations, the nucleotide sequence of the rungs on the other half of the ladder must be TTCAGC. The biological significance of this is obvious. During mitosis and meiosis the ladder separates into two halves (at the chromosomal level this is the separation of the two complementary chromosomes). In mitosis, new complementary halves of each of the separated halves are built using the 'code' contained in the original halves. The result is two strands of identical DNA: one for each of the newly created cells. This solves the mystery of the replication of DNA.

There are two kinds of cells in nature: prokaryotes and eukaryotes. Prokaryotes contain DNA but there is no nucleus in the cell. In eukaryotes, there is a nucleus in which the chromosomal DNA is contained, with some non-chromosomal DNA existing outside the nucleus. In later chapters, the importance of the difference between these cells will become a little clearer. For now, the focus is on eukaryotes since the cells of agricultural plants and animals are eukaryotes. As indicated in the previous section, in the resting phase, chromosomes exist in matched pairs (homologous chromosomes) in the cell nucleus – the number of pairs differs according to the particular species. In mitosis, the chromosomes separate and the two strands of the double helical DNA separate. A complementary strand for each single strand is then constructed resulting in duplicate homologous chromosomes. After this process of duplication, each set of homologous chromosomes moves to the opposite pole of the cell, and nuclear membranes begin to form around each set, after which the cell divides in the centre of the two poles to create two new identical cells. In meiosis, an additional division takes place without any replication. Each new cell (gamete) after this further division contains only one of the chromosomes (one half of the DNA ladder) from each homologous pair (cells with only one chromosome from each pair are called haploid). When two gametes unite (fertilisation), a new cell is formed and has a complete set of homologous chromosomes (cells with paired chromosomes are called diploid). The chromosomes in this new cell, although derived from the parent cells, are different from either parent.

The second code embedded in DNA relates to the construction of proteins. Proteins are chains of amino acids and they are the main structural material of cells and organisms (structural proteins) and the main entities involved in cell functioning (functional proteins). Structural proteins are the main elements from which cells are constructed. They, thereby, are also the materials from which parts of multicellular organisms (such as mammals) are constructed, parts such as bone, liver, muscles and blood cells. Proteins also perform many diverse functions in cells. A class of proteins called enzymes regulate cell processes; most of the essential process would not occur without their action or would occur at rates far too slow to support cell and organism life. With respect to the coding function of DNA, the important feature is that proteins are composed of amino acids. Amino acids are simple chemical compounds. All amino acids have a common structure – an amino group (two molecules of hydrogen and one of nitrogen) and a carboxyl group (one molecule of carbon, two of oxygen and one of hydrogen). They differ only with respect to a side chain (a radical group R), as shown in the diagram.

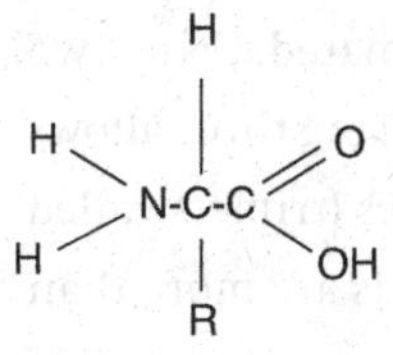

H
H O
N-C-C
H OH
R

Table 1.1 *Codon dictionary*

	U	C	A	G
U	UUU Phe UUC Phe UUA Leu UUG Leu	UCU Ser UCC Ser UCA Ser UCG Ser	UAU Tyr UAC Tyr UAA STOP UAG STOP	UGU Cys UGC Cys UGA STOP UGG Trp
C	CUU Leu CUC Leu CUA Leu CUG Leu	CCU Pro CCC Pro CCA Pro CCG Pro	CAU His CAC His CAA Gln CAG Gln	CGU Arg CGC Arg CGA Arg CGG Arg
A	AUU Ile AUC Ile AUA Ile AUG Met and START	ACU Thr ACC Thr ACA Thr ACG Thr	AAU Asn AAC Asn AAA Lys AAG Lys	AGU Ser AGC Ser AGA Arg AGG Arg
G	GUU Val GUC Val GUA Val GUG Val	GCU Ala GCC Ala GCA Ala GCG Ala	GAU Asp GAC Asp GAA Glu GAG Glu	GGU Gly GGC Gly GGA Gly GGG Gly

Twenty standard amino acids (i.e. 20 different R side chains) occur in proteins (glycine, alanine, valine, leucine, isoleucine, methionine, phenylalanine, tryptophan, proline, serine, threonine, cysteine, tyrosine, asparagine, glutamine, aspartic acid, glutamic acid, lysine, arginine and histodine). Proteins are built by stringing amino acids together. This can be thought of metaphorically as threading beads of 20 different colours together. With 20 different amino acids available, proteins comprised of 10 amino acids have 20^{10} (slightly more than 10 trillion) different possible combinations. Proteins with a string of 20 amino acids have 20^{20} possible combinations. The sequence of nucleotides on the separated ladder of DNA determines the specific amino acid to be added to the chain and the location in which it is added. Clearly, using only one nucleotide of DNA to determine which amino acid goes where is inadequate since only 4 amino acids could be designated. Using two nucleotides would allow the designation of 16 amino acids. Using three allows all 20 to be designated. And indeed, sets of three nucleotides (triplets called codons) are what evolved. Obviously, triplets of 4 amino acids are more than is needed to code 20 amino acids. Since order matters, there are 64 possible triplet combinations of 4 nucleotides. The unravelling of the code revealed that there is a lot of redundancy in the coding (there is more than one codon for all amino acids except methionine and tryptophan); also there are codons for stopping the creation of a string of amino acids and one that does double duty, coding for 'start the protein building process' and for methionine (see Table 1.1). When the codon for methionine (AUG) occurs at the beginning of the chain it codes for start, everywhere else it codes for methionine. The process of building proteins from the code embedded in DNA, unlike replication, involves another molecule RNA (ribonucleic acid). RNA is similar to DNA. One of the ways it differs from DNA is the substitution of the nucleotide uridine for thymine. Hence, when RNA is transcribed from DNA, uridine and not thymine is paired with adenine. Proteins are constructed by 'reading' triplets of nucleotides from RNA (DNA and RNA are directional with 3′ and 5′ ends, and 'reading' nearly always begins at the 3′ end); RNA is transcribed from DNA (i.e. RNA is built by 'reading' triplets from DNA). Consequently, codons are triplets of adenine (A), cytosine (C), guanine (G) and uridine (U).

2 Application of genetics to agriculture

2.1 Genetic modification of plants and animals: techniques

Modifying an organism requires altering its DNA: adding, deleting or substituting a string of nucleotides that code for a trait in the mature plant, animal, bacterium or fungus. This can be done directly or by using a vector – an entity that will modify an organism's DNA. Both methods rely on the ability to cleave (cut) DNA at desired locations and ligate (join) pieces of DNA. When a vector is used, the modification is made to the vector's DNA; the vector then modifies the organism's DNA. Use of vectors is common in plant biotechnology, as it also is in medical and environmental biotechnology that involves modifying bacteria. I discuss below the use of an element in the bacterium *Agrobacterium tumefaciens* as a vector in plant modification. A virus, λ phage, that infects bacteria is commonly used to modify the DNA of bacteria in medical and environmental biotechnology.

A number of direct modification techniques are used on animals: retrovirus-mediated transgenics, pronuclear injection (the most common), nuclear transfer to embryonic stem cells, and sperm-mediated transfer. The potential opened up by development of these techniques is impressive but, to date, GM animal agriculture is in its infancy. I set out the reasons for this below.

2.1.1 Cleaving and ligating

Fortuitously for genetic engineers, there is a class of naturally occurring enzymes that cleave DNA at specific sites (areas with specific nucleotide sequences). Two known functions of these enzymes (known as restriction enzymes) are: (1) to allow a pathogen to alter or destroy another organism's DNA, or (2) to allow an organism to defend itself against foreign DNA by being able to alter or destroy the invader's DNA. As a result, restriction enzymes are numerous and diverse. A second aspect, worth noting in passing, of the existence and functions of restriction enzymes is of less importance to human-directed genetic modification but essential for cells. Since cells produce restriction enzymes for the second function, it is important that they have a way of protecting their own DNA against the cleavage potential of the restriction enzymes they produce. This is done through a methylation system, the details of which are not important for understanding genetic engineering.

The first restriction enzyme was isolated in 1968 from the bacterium *Escherichia coli* (*E. coli*). *E. coli* is named after the German physician Theodor Escherich, who discovered it. It has been extensively studied and has been widely used in medical and environmental biotechnology (to produce, for instance, pharmaceuticals, and to degrade spilled oil). In these contexts, it has many advantages. For example, it is easy and inexpensive to grow (it has a rapid doubling time: 20–30 minutes), laboratory strains contain mutations that make survival outside the laboratory impossible, and it contains DNA outside its chromosomes (extra-chromosomal DNA), which can be used as vectors. Unfortunately, its restriction enzyme, *E. coli* K, has complex characteristics, which render it difficult to study and use. Its discovery, however, initiated the quest for other restriction enzymes, and two years later a much more useful restriction enzyme was isolated from the bacterium *Haemophilus influenzae*, the restriction enzyme *H. influenzae* Rd. This enzyme cleaves the DNA of a bacteriophage (T7), a class of viruses to which I shall return later.

Restriction enzymes cleave DNA at specific sites known as their **recognition sites**. A recognition site is a small segment of complementary strands of DNA. In the case of *H. influenzae* Rd (also designated HindIII), the nucleotide sequence at the recognition site is:

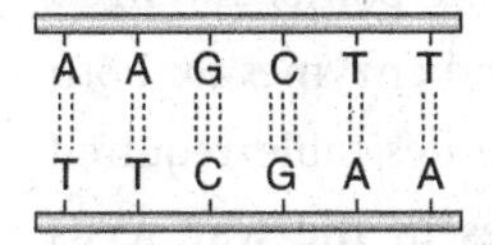

This recognition site is six nucleotides in length. The number varies with the restriction enzyme. This restriction enzyme breaks the rungs of the ladder at this six-nucleotide location, separating the strand of DNA.

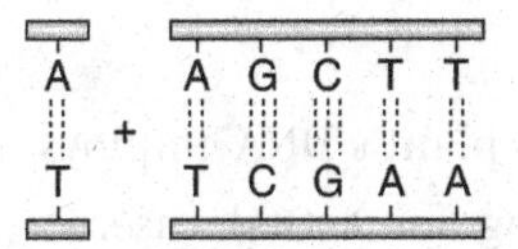

Since the two strands are complementary, only one need be specified in this case, but a feature of the most useful restriction enzymes does require a specification of both strands. In 1972, the restriction enzyme EcoRI was isolated from the RY strain of *E. coli*. Its recognition site is:

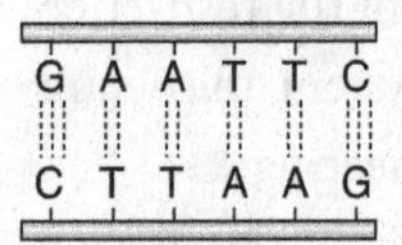

What makes this enzyme, and others like it, interesting and important in genetic engineering is the nature of its cleavage pattern. Instead of cleaving DNA at the opposite ends of the recognition site, it cleaves DNA some place in the middle of the recognition site. Specifically, in this case, the cleavage pattern is:

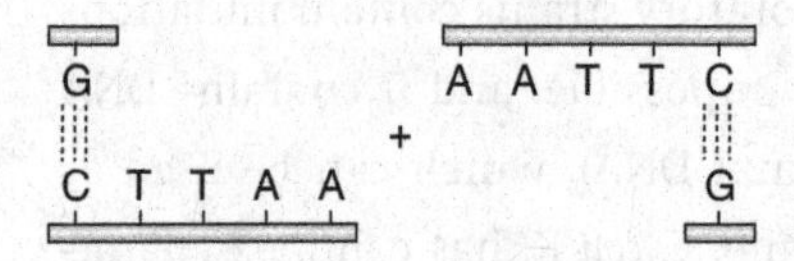

This pattern, termed 'cohesive ends', or colloquially, 'sticky ends', is important because the exposed single strands make ligation easier.

When two complementary sticky ends meet (ends with complementary base pairing), they associate – weakly join together. To complete the joining requires that a continuous sugar-phosphate backbone be formed. This requires another enzyme, DNA ligase. This enzyme catalyses the formation of a phosphodiester bond between two DNA chains; its essential role, in nature, is to repair nicks in DNA, but in genetic engineering it is used to ligate a human-introduced strand of DNA to an existing strand.

Techniques for cleaving (separating) DNA at appropriate points and ligating (joining) strands of DNA are now well understood, and enzymes for both processes are available to biotechnologists. Most of the desirable required enzymes can be purchased from specialised companies in the way seeds can be purchased from companies that specialise in seed development and production.

2.1.2 Vectors

As indicated above, a common method of modifying a plant's DNA employs a vector. This can be easily explained by describing an actual case. A

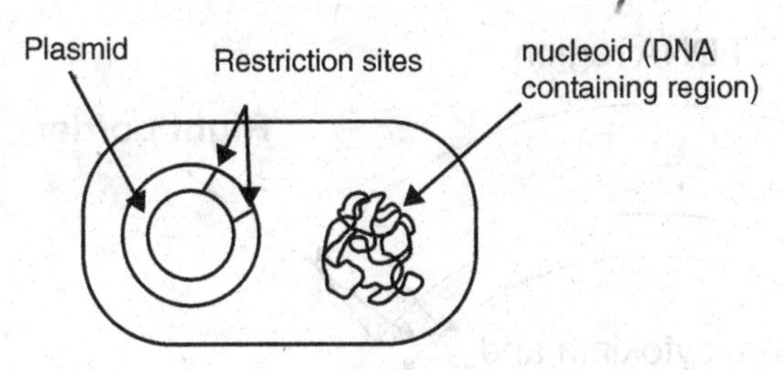

Figure 2.1 Bacteria are prokaryotes; they do not have a nucleus but do have a nucleoid composed of DNA that functions in the same way as that in the nucleus of eukaryotes.

common bacterium that is ubiquitous in soil, *A. tumefaciens*, causes a tumour-like growth on plants called crown gall. The bacterium seldom affects healthy, uninjured plants since its usual point of entry is a break in the cell wall. *A. tumefaciens* contains the Ti plasmid (Ti = tumour inducing). Plasmids are doubled-stranded lengths of DNA, usually circular, and occur almost exclusively in bacteria. Bacteria are prokaryotes; hence, their chromosomes are not enclosed in a nucleus. Plasmids are non-chromosomal (designated extra-chromosomal) DNA, which replicate independently of the bacterium's chromosomes. Chromosomes, whether enveloped in a nucleus or not, carry the genetic information of the cell and, in multicellular organisms, the organism comprised of those cells; chromosomes carry the code for constructing the cell and its processes. Plasmids have specific functions in the cell but do not carry the cell's genetic information. The Ti plasmid of *A. tumefaciens* (see Figure 2.1) is a circle of double-stranded DNA. It is a large plasmid consisting of about 200,000 base pairs (bp) (rungs on the ladder). The length of DNA is expressed in terms of the number of base pairs, the thousands being represented by *k*. Hence, the Ti plasmid is 200 kbp.

A section of the plasmid codes for tumour production (see Figure 2.2). It is about 23 kbp in size. Two other regions of the plasmid are critical to its action: virulence genes and opine utilisation genes. Once a bacterium enters a plant cell, these genes integrate the tumour (T)-DNA into the chromosomal DNA of the plant cell. The altered chromosomes result in cell reproduction that forms the tumour. It is this integrating (insertion) process that makes this plasmid especially useful in genetically modifying a plant.

To use this plasmid as a vector for genetic modification, we need to modify it. The first step is to remove the plasmid from the bacterium. The next is to cleave the plasmid's DNA, using a restriction enzyme, at the beginning

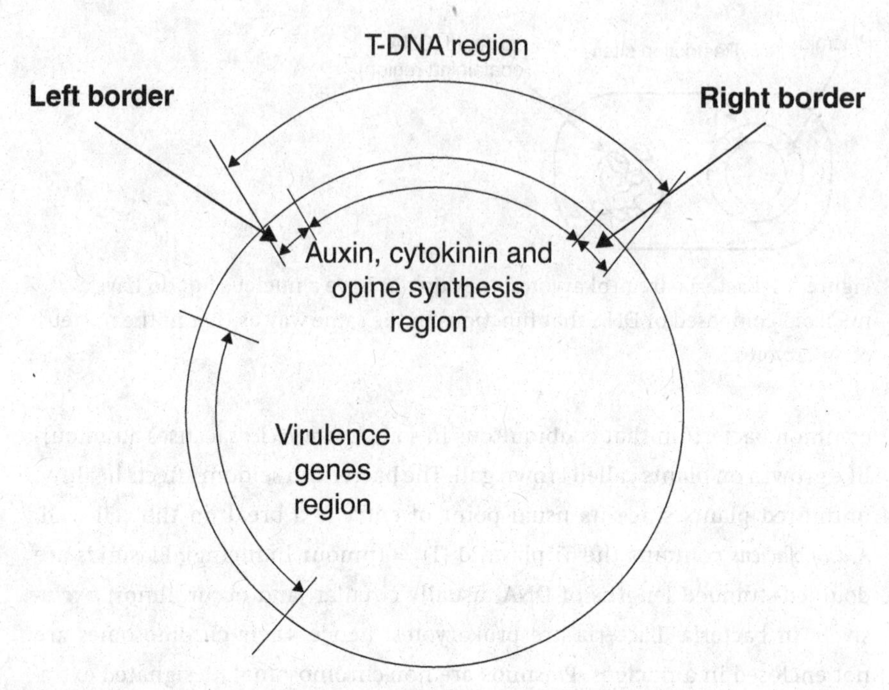

Figure 2.2 Regions of the Ti plasmid of *Agrobacterium tumefaciens*.

and the end of the T-DNA segment. This segment is then removed (thereby disarming its tumour-producing capacity). In its place, a segment of DNA that codes for the desired trait – such as a glyphosate (Roundup)[1] resistance trait – is inserted and joined to the free end of the cleaved plasmid with DNA ligase. The plasmid still retains the segments of DNA required to insert the target DNA (now located where the T-DNA used to be) into the chromosomal DNA of the plant. The plasmid is then reinserted into the bacterium. Leaves of the plant to be modified are damaged – often by making holes with a hole punch. This allows the bacterium with its modified Ti plasmid to infect the plant cells. The plasmid will integrate the DNA that has replaced the T-DNA into the chromosomal DNA of the infected plant cells. After this, the chromosomal DNA of the plant cell will contain the target DNA. The modified cells are then induced to undergo embryological development, which will result in a

[1] Roundup is a Monsanto trademark. Until 2004, Monsanto held the patent on glyphosate (the active ingredient in Roundup: an alternate name for this ingredient is phosphinothricin). The generic name is glyphosate. It is a broad-spectrum herbicide, meaning it will kill plants indiscriminately. Genetic modification of crops can make them resistant to glyphosate. Monsanto calls these modified crops 'roundup ready'. This genetic modification will be set out in more detail later.

plant with the agriculturally useful trait. Since it is the chromosomal DNA of the plant that has been modified, any seeds it produces will carry the gene for the trait. This is a potent and highly efficacious method of modifying a plant's DNA. This vector can be readily modified outside the bacterium and reinserted. Subsequently the bacterium and its Ti plasmid do all the genetic work of modifying the plant's chromosomal DNA.

2.1.3 Pronuclear microinjection and cloning

Two methods of microinjection have wide use in genetically engineering animals. Cloning animals (e.g. Dolly) involves microinjection directly into the nucleus of the cell of the animal. Cloning requires evacuating the nucleus – taking out the original DNA – and injecting into the cell the DNA of the animal being cloned. The other method injects a pronucleus. Immediately after a sperm cell and an egg cell join (fertilisation), the nucleus of each is separate (pronuclei of what will become a new fused nucleus). Foreign (target) DNA is injected into one of the pronuclei before fusion. This pronuclear injection process is performed on numerous egg cells. These egg cells are then transferred to a surrogate mother. Between 10 and 30 per cent of the eggs will contain the foreign DNA. Animals (the preferred experimental animal is the mouse) expressing the foreign DNA are bred. Genetically modified agricultural animals play an insignificant role in food production. In the next section, I give examples of a medical and an industrial use of goats but neither connects with common agricultural production or products. There are other methods of inserting DNA into the nucleus such as gene guns. These other techniques are used more frequently in medical contexts.

2.2 Agricultural biotechnology: current products and future prospects

As stated in Section 2.1, genetic modification of agricultural animals is modest. The most prominent arena of modification uses animals as biofactories; animals are modified to produce a valued product. Even though the animals are domesticated farm animals, such as goats, the products are not related to food agriculture. These modified farm animals do not increase agricultural productivity or the quality of common agricultural products. Instead they are genetically modified to produce medically and industrially useful products.

For example, goats have been genetically modified to produce in their milk a form of human antithrombin, which is used in the treatment of individuals with hereditary antithrombin deficiency, a disorder which has a prevalence in the human population of 0.2–0.4 per cent. The modification was developed by GTC Biotherapeutics and is marketed by it under the registered name ATryn. In August 2006, the European Commission, and in February 2009 the Food and Drug Administration in the USA approved ATryn. This is noteworthy because it is the first pharmaceutical produced by genetically modified farm animals (goats) to be approved for clinical use. Goats have also been genetically modified to produce in their milk an industrially useful fibre. Nexia modified goats and marketed the fibre under the registered name BioSteel. BioSteel is spider silk (a protein used by spiders for their web construction); its value lies in its strength, lightness and flexibility.

There are several reasons for the modest developments in creating transgenic animals whether for medical, industrial or agricultural use. First, the genetics is more complicated; the behavioural, anatomical and physiological characteristics of agricultural animals are quantitative traits. Multiple genes, complex development processes and numerous environment factors are involved in introducing a novel trait or enhancing an existing trait. Many beneficial traits can be introduced into plants because the nucleotide sequence for the trait is relatively small and all in one location; ideal vectors exist, such as *A. tumefaciens*; and the developmental pathway from the nucleotide segment to the expression of the trait is reasonably straightforward. This is not so in the case of farm animals because, as indicated, most of the important traits are quantitative traits; there are multiple, non-contiguous, interacting strands of DNA and there are complex regulatory and embryological developmental processes. Not surprisingly, the genetic modifications of animals involve single-location, short nucleotide sequences that involve a reasonably simple developmental process.

A second impediment to genetically modifying agricultural animals centres on the complexity of the techniques involved and the success rates of the outcome. The ova and sperm have to be removed, manipulated *in vitro* and then transferred into an animal for gestation. The two common procedures – cloning and pronuclear injection – require considerable skill. Cloning requires the removal of the nucleus of a fertilised ovum (zygote) and replacing it with the nucleus from a somatic cell. At the blastocyst stage (about 7–8 days after fertilisation in humans and cows), the embryo is transferred to a host animal.

Pronuclear injection requires the injection of foreign DNA into one of the pronuclei before the female and male pronuclei (ovum and sperm) fuse. Upon fusion of the two pronuclei, the foreign DNA becomes incorporated into the chromosomal DNA of the zygote. The embryo is then transferred to a host animal. Even though the biological processes are well understood and the techniques well honed, the success rates – live births with the intended DNA – are low. The use of stem cells may increase the success rates but the techniques are in the early stages of development.

Another challenge is commercial scale-up; that is, taking laboratory-scale success to industrial-scale production. Eyestone (1998) has remarked, 'A primary aim of transgenesis is to establish a new genetic line in which the transgenic modification is stably transmitted and expressed both within and between generations.' Currently, establishing genetic lines is expensive and the success haphazard. Commercial scale-up depends on developing a large population within an economically feasible time frame; the long gestation periods and small numbers of offspring per pregnancy of most agricultural animals significantly extend the time to commercialisation. Plants, which produce large numbers of seeds per plant, are more economically feasible. The scale-up time for animals being used as biofactories is longer than for plants but substantially less than that for modifications for enhancing productivity or quality of common agricultural products. One goat can produce large quantities of antithrombin or spider silk in its milk; hence, a small herd can produce financially adequate amounts of the product.

Consequently, the greatest impact of genetic modification in agriculture has been on crop agriculture; in terms of hectares planted and annual commercial value, corn, soybean, canola and cotton are at the top of the list. Many other plants have been genetically modified, such as wheat, rice, potatoes and eggplant. There has been a spectrum of properties molecularly inserted into agricultural plants – vitamin enrichment as in golden rice and seedless fruit such as in eggplant, for example – but the central focus for major crops has been on pests and weeds. Modifying plants to make them resistant to the herbicide glyphosate (*N*-phosphonomethyl glycine – commonly marketed under the brand name Roundup) is the dominant GM strategy for weed control. Modifying them so they produce a toxin is the dominant GM strategy for pest control. The immediate beneficiaries have been farmers, for whom yields have gone up and some input costs have gone down; benefit to consumers, if any, would relate to food costs.

Let's look first at glyphosate resistance. Glyphosate is a non-selective, broad-spectrum herbicide; it kills almost all plants. It is absorbed through the leaves and moves via phloem to the roots (phloem is a tissue in the vascular system of plants that transports organic material within the plant). Glyphosate inhibits the action of an enzyme, 5-enolpyruvoylshikimate-3-phosphate synthase (EPSPS), found in plants and some microorganisms. EPSPS is essential to the production of EPSP through the shikimic acid pathway. The next step in the pathway is chorsimate production; chorismate is essential to the synthesis of aromatic amino acids (phenylalanine, tyrosine and tryptophan). As you will recall, amino acids are the building blocks of proteins, so by inhibiting the synthesis of essential amino acids, the ability of the plant to engage in cellular activity is undermined and the plant dies. Sikorski and Gruys (1997) provides an excellent detailed account of this process.

As discussed more fully in Chapter 5, the immediate benefits to farmers of glyphosate are fewer herbicide applications and zero tillage (i.e. the soil does not need to be tilled – ploughed or harrowed – prior to planting a crop).[2] Tilling breaks up the soil and, among other things, disrupts plants and seeds; this kills existing plants and retards the emergence of new ones. The field is then planted with the crop; the goal is to have the crop emerge before (or at least simultaneously with) the emergence of a new crop of weeds. Without tillage, the emerging crop plants struggle for survival (for nutrients, sunlight, space and so on) against the weeds. The disadvantage of tilling is that it exposes the soil to wind and water erosion, which over time will decrease the fertile topsoil, and it often increases water loss through evaporation. Glyphosate can be applied to the field prior to planting. Usually, in 4–7 days, the weeds (including shrubs) will wither. The crop can then be planted. This, of course, does not require glyphosate-resistant crops; seeds of a conventional crop can be planted after spraying with glyphosate because it is absorbed through the foliage, and there will be no foliage for many days and, for some plants, many weeks. Glyphosate resistance allows subsequent sprayings of glyphosate – that is, spraying after the crop plant has foliage and weeds have reappeared. The glyphosate-resistant crops will not be affected by glyphosate but all the weeds

[2] There are numerous methods of zero tillage: mowing or rolling existing weeds at the end of a growing season and planting a cover crop, or covering the field with a light, impervious fabric, cardboard or other material, for example. Glyphosate requires fewer steps and is more predictable. Nonetheless, it does require the application of a chemical while some other methods do not.

will. Sometimes a single, well-timed, mid-season application is effective for weed control through to harvest. Moreover, because glyphosate – along with some other herbicides such as Linuron and Imazapyr – is a non-specific, broad-spectrum herbicide, it can be used alone for control of all weed types. Many other commonly used agricultural herbicides are specific. For example, 2,4-D and Profluralin are most effective for broad-leaf plants, whereas Metolachlor is most effective on annual grasses. To obtain a broad-spectrum effect, herbicides that are specific to certain types of plants have to be combined, or more than one spraying is required.

Turning now to *Bt* crops, these have been genetically modified to express δ-endotoxins (Cry toxins) which are toxic to the larvae of moths, butterflies, flies, mosquitoes, weevils and beetles. The toxicity is specific to the larval stage of these organisms (in the case of corn and cotton, it kills the larvae of the boll weevil, corn rootworm and European corn borer). There, however, have been reports of some Cry toxins killing wasps and bees (Garcia-Robles *et al.*, 2001) and nematodes (Marroquin *et al.*, 2000; Wei *et al.*, 2003). In nature, these δ-endotoxins are produced by the bacterium *Bacillus thuringiensis*, which occurs naturally in soil. The nucleotide sequence inserted in *Bt* crops is derived from *B. thuringiensis*. Spraying the bacterium on crops as a pesticide has been practised for at least 50 years; in most jurisdictions it is considered an 'organic' pesticide and has been, and continues to be, widely used by organic farmers.

The δ-endotoxins are crystal proteins that require proteolytic processing in the midgut of the larva to become toxic. The crystals dissolve in the alkaline environment of the larval midgut; the first step in processing. Cry crystals, as with other proteins, have restriction sites (cleavage sites). Protease is a restriction enzyme found in the larval midgut that cleaves (breaks apart) the crystal. After proteolytic processing by larval midgut protease, the toxin binds to midgut receptors on the lining of the gut, where it forms channels in the apical membrane. The channels allow water and ions to freely enter the cells, causing them to swell and then rupture (lysis). Ultimately, the larva dies. Since the toxicity of the δ-endotoxins depends on protease processing in a specific environment, their toxicity is extremely limited, making a very safe pesticide. Pigott and Ellar (2007) provides a detailed exposition of this process.

The benefit of *Bt* used as a spray by farmers is obvious. It is a safe, effective pesticide for the control of the target larvae that inflict significant damage on certain crops. A limitation, of course, is that it only affects larvae that ingest it;

hence, the rootworm bore, for example, will be unaffected. Farmers also bear the cost (time, equipment and fuel) of the spraying. GM crops that express the *Bt* endotoxin do not have the limitation and do not impose those costs, increasing significantly the benefit to farmers.

Research and development on the next generation of GM crops is focusing on traits such as drought tolerance, reduced nitrogen requirements, and the addition or enrichment of health-promoting compounds (such as soybean and canola oils that contain long-chain Ω-3 fatty acids, which promote cardiac health and which currently are derived principally from fish oils). The potential benefits of these crops extend beyond farmers; they have important environmental amelioration and sustainability impacts, positive human health impacts and positive impacts on poverty in low- and middle-income countries. There are, of course, concerns about attendant harms. The benefits are discussed in detail in Chapter 5 and the harms in Chapter 6.

3 Philosophical and conceptual background

Consider the claim, 'Inserting a gene into a plant that causes it to produce a compound that is toxic to the larval stage of many insects is harmful and should be prohibited.' This, on the surface, seems to be a straightforward claim, but below the surface lurk numerous logical inferences, presuppositions and evidential claims. The evidential claims on which it rests are the most obvious of these. Its acceptability depends on there being evidence that genes result in certain traits, that a specific compound is toxic to the larval stage of specific insects, and that this leads to consequences, such as the safety of food, the mortality and morbidity of people, and/or the integrity of ecosystems and the environment.

It also rests on an assumption that these consequences are negative (harmful). This might be taken to be another empirical (scientific) assumption. In part it is, but it goes well beyond empirical evidence and imports values. A change in an ecosystem, for example, is not in and of itself negative or harmful. Were ecosystems not constantly in flux – a constant flux that was occurring for more than a billion years before the emergence of hominids – the evolution of the rich biodiversity of life today would never have occurred, and hominids would not have emerged. Deeming consequences negative is inextricably connected to human goals, desires and ideas. In most value (ethical) systems, unsafe food, increases in human mortality and morbidity, and a decrease in biodiversity or the stability of ecosystems due to a human activity are harms because of their effect on humans. But even in those few systems where advocates maintain that humans have a moral obligation that transcends their own welfare to minimise the impact of human activity on nature, a value is being appealed to – perhaps given by a deity in a command to be a faithful steward of the Earth or some such external source.

Moreover, the original claim rests on logical inferences. There is an inference, for example, from harm to prohibition. A presupposition such as, 'If something is harmful, it should be prohibited', is required for this inferential transition to be valid. Once exposed as a required presupposition, it needs to be justified. There are lots of harms that we tolerate or are willing to risk occurring because a greater good is perceived to be connected to the activity. There is no shortage of harms and potential harms for a woman associated with pregnancy and childbirth, but it is hard to imagine anyone advocating that pregnancy be prohibited. Similarly, there is no shortage of harms and potential harms associated with automobile use. There are significant questions about how we use automobiles – overuse, fuel-inefficient vehicles, patterns of use and so on – but very few would support a prohibition based on the existence of harms.

What does occur when harms are identified or there are grounds for concern about potential harms is that the harms and benefits are identified and weighed, and a rational decision about the most prudent and defensible course of action is made. This is known as risk analysis and is the basis for rational decision-making involving harms and benefits. Benefits can seldom be realised without some harms or risk of harm.

So, the simple claim, 'Inserting a gene into a plant that causes it to produce a compound that is toxic to the larval stage of many insects is harmful and should be prohibited,' requires an examination of evidence, values and reasoning. Engaging in these is at the heart of philosophical and logical analysis, an activity that is far from trivial, is complex and extends far beyond agricultural biotechnology. Understanding the elementary logic of reasoning, the ethical theories from which values, and value claims, are derived and the rational techniques for balancing and managing harms and benefits is essential to navigating the landscape of agricultural biotechnology – and, it is worth adding, nearly every other aspect of human activity and decision-making. In this chapter, these aspects of philosophical analysis are sketched in sufficient detail as to allow the reader to appreciate the techniques of analysis employed in Chapters 4–8.

The major ethical theories that frequently lie beneath the surface of claims made in the controversies over agricultural biotechnology are also described. One of the conclusions of Section 4.2, where the moral (often theologically based) objection to inferring with life (playing God) is examined, is that even those who hold ethical views that are based on fundamental ethical principles cannot avoid the need to engage in harm-benefit analyses. Hence, rational assessment of harms and benefits is essential regardless of what ethical theory one adopts and brings to the discussion. The elements of the risk assessment will be different for those holding different ethical theories but I contend the outcome will be much the same. This might make some readers wonder why space is devoted to setting out ethical theories when, ultimately, which one is held matters little to resolving the issues involved in agricultural biotechnology. For me, there are three reasons. First, I can only make the case that the ethical theory adopted is less relevant than might initially appear if a reader knows what the alternative theories are and why some people might think they make a difference. Second, a reader is only in a position to decide whether my claim is credible if the competing ethical theories are understood in sufficient detail. Third, the actual process of reasoning and the value assumptions made will differ for advocates of different theories. I contend that the paths followed within different ethical traditions converge in the decision-making outcome in the context of agricultural biotechnology; that does not make the paths themselves irrelevant to those who want to follow them to reach a decision.

3.1 A primer of logic, reasoning and evidence

This section has two goals. First, it sets out the tools that I will use in a variety of contexts in subsequent sections. Hence, a reader can see the role of the evidence cited and the reasoning processes employed. Second, and for me the more important goal, is to arm the reader with tools of analysis that will enable her to dissect my claims, reasoning and purported evidence. This maximises the probability that the reader is intellectually engaged in a dialogic process and that conclusions reached are not simply based on persuasive language but on critical examination (analysis) of the material, evidence and reasoning.

Logic, both symbolic (mathematical) and informal, codifies rules of thought or reasoning; rationality involves following those rules. Reasoning draws on two quite different kinds of logic: deductive and inductive logic. Inductive logic involves inferring a general statement from specific instances, such as inferring, 'water expands when it reaches a few degrees below 0°C' from many observations of this phenomenon. Deduction is inferring a claim from other accepted claims and is the logic principally used in this book. Induction, however, is a mode of inference essential to scientific reasoning; it is used in confirming hypotheses, forming generalisations about regularities and constructing theories. Hence, it will be relevant at various points. Mostly, I, as is common throughout science, will use peer review and, more generally, scientific community acceptance as a guide to the reasonableness of accepting scientific claims; this strategy, though not infallible, has proved over time to be extremely robust. The ideal, obviously, in maximising one's confidence in some purported evidence, is to examine every piece of relevant research oneself and reach a judgement but this is a daunting task. Hence, again, as is common, I do this only in cases where there exists a debate within the scientific community or a deeper understanding of the evidence is required. Those who are interested in delving further into inductive logic in science should consult the excellent treatments by Colin Howson (2000), Brian Skyrms (1966) or Von Wright (1960).

To avoid confusing readers who have encountered induction in mathematics, it should be noted that induction in mathematics and induction in logic are different methods. In mathematics it is, actually, an instance of logical deduction. Proving the truth of every proposition in an infinite series of propositions cannot be done by proving the propositions one by one; after all, the series is infinite. Mathematical induction is the technique of proving that the first proposition in the sequence is true and also proving that if any proposition in the series is true, the next one is true as well. Consequently, they must all be true.

Several crucial assumptions underlie elementary mathematical logic (also known as first-order predicate logic with identity). First, a claim (*A*) cannot be both true and false. Second, either a claim or its negation (*A* or not *A*) must be true; they cannot both be false (the principle of excluded middle). Third, either a claim or its negation must be false; they cannot both be true (the principle of non-contradiction). There are abstract logics in which one or more of these assumptions are denied. In many-valued logics, for example, the principle of excluded middle is not assumed; claims do not have to be true or false since there are other possible 'values'. In a logic that Graham Priest (1986) has dubbed 'dialetheism', the principle of non-contradiction is not assumed; a claim can be both true and false. Modal logics introduce the concepts of necessity and possibility, which transform the understanding of the three principles. This plethora of alternate logics is similar to alternate geometries (Euclidean, spherical and hyperbolic, for example). Only one is held to describe the actual nature of space. Until Einstein, the one that was accepted as describing space was Euclidean; Einstein's general theory of relativity assumes space to be best described by spherical geometry. Similarly, of the many logics, elementary mathematical logic is taken as the logic of thought and reasoning.

In setting out these assumptions, I have used the terms 'true' and 'false', mostly because they historically have been expressed using these terms. There is a rich literature devoted to specifying what 'true' means. For our purposes, 'true' simply means 'accepted'. There is no need to open up the, probably irresolvable, debate over the meaning of 'true' and the grounds for claiming something to be true. The evidence on which one accepts a claim is, of course, critical but a decision about whether accepting a claim given the evidence is reasonable does not require employing the term 'true'. Everyone holds certain claims to be sufficiently probable, given the evidence, to warrant their acceptance and that is enough. The total collection of claims a person accepts constitutes that person's conceptual framework. Some claims in that framework will be accepted on the basis of empirical evidence, a feature examined later in this section. Some claims, specifically for religious individuals, will be accepted on the basis of theology, revelation (such as scriptural truths or disclosures from a spiritual realm), or natural reasoning (such as truths gleaned from the handiwork of God as seen in organisms or the night sky). Some claims will be accepted because they can be deduced from other claims. Deduction is the application of rules of reasoning – logic. Some claims will be accepted because they can be induced from other claims. Induction is also the application of rules of reasoning. Deduction is more robust and reliable than induction.

Deduction is the process of connecting claims to each other in ways that make acceptance of a deduced claim **necessary *if*** the claims from which it is deduced are accepted. As a result, at the heart of deduction is 'truth/acceptance preservation'. Deduction appeals to rules that are truth-preserving; only rules that connect claims in ways that make **necessary** the acceptance of one claim on the basis of one or more other accepted claims are legitimate. The claim that is being deduced is known as the conclusion; the claims from which it is being deduced are known as premises. The rules of deductions ensure that a conclusion cannot, rationally, be rejected (i.e. not accepted as true) if the premises are accepted. A collection of premises, a conclusion and relevant rules of deduction are known as an argument. The term, 'argument', in logic, is a technical term and has nothing to do with its colloquial use to designate a disagreement. So, we have premises from which conclusions can be deduced and together they constitute an argument; hence, to offer an argument for a claim one is advancing requires the provision of some premises from which it can be deduced. Arguments that employ only legitimate (truth-preserving or, more broadly, acceptance-preserving) rules of deduction are known as valid arguments. A final, but exceptionally important point about arguments and deduction is that valid arguments only guarantee that the conclusion must be acceptable ***if*** the premises are acceptable; it does not guarantee that the premises are, in fact, acceptable (justifiable). Hence, the process of reasoning requires two things: that the argument advanced be valid and that its premises be acceptable. Arguments that are valid and have true (accepted) premises are known as 'sound' arguments. This is why evidence is central to 'sound' reasoning.

With this understanding of deductive reasoning, let's return to a more detailed examination of the logical requirements for coherent conceptual frameworks. Again, as a matter of elementary logic, any collection of claims a person holds as true (accepts) must be consistent; hence, one's conceptual framework must be consistent. A set of claims is inconsistent if, using truth-preserving rules of deduction, a contradiction can be deduced (that is, both a claim and its negation can be deduced from the same set of claims). That a contradiction can be deduced indicates that one or more of the claims comprising the conceptual framework is false. Although there has been much discussion among logicians about whether the existence of a contradiction in one's conceptual scheme is fatal, for the most part, it is undesirable and a sign that trouble is afoot. The reason is elementary. If claims *A* and not-*A* are both accepted, then in any case where claim *B* is derivable from a set of claims containing *A*, the claim not-*B* can be derived from a set of claims containing not-*A*. In short, every claim and its negation can be deduced.

A method of proof in logic, known as *reductio ad absurdum* (reduction to the absurd), is based on the principle of non-contradiction; if assuming a claim allows a contradiction to be deduced, then the claim has reduced the framework that contains it to absurdity (a contradiction being an absurdity). In mathematics, *reductio ad absurdum* proof is known as 'indirect proof'; it also relies on the fact that if a contradiction can be deduced from a set of propositions, one of the propositions must be false. This method of proof is also contextual; all the propositions in the set except for the proposition that one wishes to prove must be accepted as true. If assuming the negation of the proposition to be proved leads to a contradiction, the negation must be false and the original proposition must be true.

Inconsistency ultimately rests on the existence of contradiction but frequently the contradiction is mediated. The system is inconsistent because two claims are contraries rather than out-and-out contradictions. Two claims are contraries if they cannot both be true even though they could both be false. For example, 'I am in a movie theatre' and 'I am standing in the middle of a highway'. They cannot both be true but they could both be false; I might be in a hot air balloon high above the Serengeti plains. The existence of contrary claims is all that is needed to make a set of claims inconsistent. That, however, is because contraries allow the deduction of contradictions. The claim, 'I am standing in the middle of a highway', entails the claim, 'I am not in a movie theatre', and hence a direct contradiction can be generated. Inconsistent sets of claims often go undetected because the offending claims are contraries rather than immediate and explicit contradictions. One feature of analytical examination of a set of claims – or a few claims in the context of the set – is to flush out contraries since they will always entail a contradiction.

With some esoteric exceptions, philosophers and mathematicians have insisted on internal consistency of any formal system – conceptual frameworks, including theories in science and ethics, are formal systems in the mathematical sense. Some philosophers/logicians have suggested ways to block the deduction of all claims once a contradiction is accepted but the circumstances are highly specialised and not found in ordinary reasoning and mathematical proofs. Kurt Gödel, in a famous proof, demonstrated that no formal system can be both consistent and complete. Completeness means that from the axioms (the fundamental statements which must be assumed) every other statement within the domain of that system can be generated. So, Gödel demonstrated that formal systems that are consistent cannot be complete and vice versa. Completeness bought at the price of inconsistency is too expensive; better to be incomplete but consistent where rational thought and proof are concerned.

The consistency requirement extends beyond the internal consistency of a framework. It also entails that once an individual or society has adopted a conceptual framework, it must be used for all deliberations. Switching frameworks in order to get the desired answer is illegitimate. That is, it is illegitimate to adopt one framework to generate one claim (a claim related to food labelling, for example) and a different framework to derive another claim (a claim about public health regulations, for example). Sometimes two frameworks that appear on the surface to be different can be shown to be equivalent, or it can be shown that one is a part of the other (subsumable under the other). In these cases, one is not really switching theories but choosing the most felicitous way in which to make the case. The onus of establishing equivalence or subordination rests with individuals or a society that wishes to employ what, on the surface, appear to be different frameworks.

In addition, the consistency requirement entails that principles derived from a theory must be uniformly applied. Take the principle of deterrence, for example. Assume that a theory entails this principle: a principle of the form, if a particular penalty applied to one or more individuals results in a reduction in an undesirable behaviour, then those individuals should be penalised. Hence, if executing a murderer results in a reduction in murders, murderers should be executed. Many people embrace this principle. There are a few factual claims that require verification such as whether executing murderers really does reduce murder rates and that a specific individual is in fact guilty of murder. These, in fact, are far from empirically settled matters. Most researchers who have studied this issue find the evidence that execution is a deterrent to be weak at best; many have concluded that the data do not support a claim of deterrence – data such as the lack of difference in murder rates in adjacent, and socially and demographically similar states of the USA, one of which has capital punishment and the other not. Also, while reasonable confidence in the judicial system seems warranted, the large number of recently discovered wrongful convictions (mostly as a result of DNA evidence) suggests that individuals are frequently convicted of murder and later exonerated. But, for the purpose of this example, let's assume that these facts are supported by evidence.

Now consider a different case. A professor, for several years, has taught a course on social issues and has been told by many students that they took the course because it was considered a 'bird course' (an easy, low-standards course). This alarms her since the topics are important (discrimination, access to health care and so on) and the techniques of critical analysis and the examination of empirical evidence require well-honed analytical and reasoning skills. Senior colleagues, whom she believes have a wealth of experience, have told her that a high failure rate will deter students from this attitude and from taking the course for inappropriate reasons. So, she marks the first essay assignment and is quite impressed with the quality of the work but gives the best essay a C grade and fails 55 per cent of the class. When challenged by the students, she responds, 'I have penalised you to deter future students from holding inappropriate attitudes about the course and, thereby, acting in inappropriate ways.'

One would expect protest from the students; they can be expected to claim that this is 'unjust'. The argument they could advance would be that they should get the marks they deserved and not be *mere means* to achieving an end. Of course, the professor could remind them of the principle: if a particular penalty applied to one or more individuals results in a reduction in an undesirable behaviour, then those individuals should be penalised. Some students might even have espoused that principle in the case of murderers, in which case there is an inconsistent application of the principle at work here. One attempt to resolve the inconsistency might be based on the concept of the 'just punishment'; death is the just punishment for murder. By contrast, it is not just punishment – or just anything – to base students' grades on anything other than the quality of their work. But then the issue becomes, does the ethical theory employed entail that death is a just punishment for murder?

For the sake of discussion, let's assume it does. In this case, the principle of deterrence has nothing to do with the justice of executing a murderer; justifying that action requires the derivation of the principle of just punishment from the ethical theory. If executing murderers is just according to the ethical theory, then any deterrent effect is a fortuitous secondary effect of applying that principle. The deterrent effect is not itself the **justification** for executing murderers.

In brief, if the principle of deterrence is a fundamental or derived principle, it must be applied consistently, and this, as we have seen, requires that it be applied in cases in which it appears patently unjust – such as the students' case. So, we have a contradiction. Hence, by *reductio ad absurdum*, it cannot be a principle of justice; furthermore, in light of this contradiction, if it is a fundamental principle of an ethical theory or is derivable from the theory's fundamental principles, there must be contradictory claims among the fundamental principles, a discovery that casts doubt on the ethical theory. We will refine this criterion later; for now, this will make the point that consistent application of a principle is essential, and when an inconsistent application is detected, it must be dealt with. Dealing with it usually requires fancy logical or evidential footwork. Sometimes terms are redefined, principles are clarified (reinterpreted) and the like. How successful such moves are depends on how contorted the theory becomes under the adjustments. Minor adjustments might rescue the theory from absurdity. Significant contortions – some rendering the theory unrecognisable – perpetuate absurdities and appear to be acts of desperation on the part of irrational advocates.

Another requirement of reasoning (rational thought) is that concepts must be clear (unambiguous). They must also, as already indicated, be used consistently. Concepts such as 'discrimination', 'life', 'harm', 'interests', 'equality', 'consent' and so on are pre-analytically (i.e. before analysis and clarification) exceptionally vague. Take the concept, 'being alive'. It sometimes means physiologically alive (alive as opposed to dead); it sometimes means experientially alive (as in the claim, 'she lost three years of her life in a prisoner-of-war cell'); sometimes it means imbued with vitality (as in, 'she was alive with ideas' or 'he came alive when on stage'). The problem is not with words having multiple meanings; a problem arises, however, when different meanings are used in the same context. In discussions of abortion, appeal to 'the right to life' must employ a single meaning of 'life'; otherwise confusion ensues. More relevant to biotechnology, consider the debates over labelling food. The central justification given for demanding that food containing GM ingredients have that fact on the label is that consumers have a 'right to be informed'. Who could disagree? What it is to 'be informed', of course, is complicated (see Castle, 2006). An obvious meaning in this context is that being informed is having access to a fact. Labelling meets that meaning.

Few, however, really take 'being informed' to mean just having access to a fact. There is a reason why, in a court of law, individuals are asked to swear to tell the truth, the whole truth and nothing but the truth. Facts are contextual and words are open to misinterpretation. Partial information may be true but misleading. A patient is 'being informed' when a physician truthfully tells him that in 74 per cent of people who have used pharmaceutical X, it has been successful. But if the physician means by 'successful' that the person lived five years longer than he might have otherwise, that might not be what the patient assumes to be success. Moreover, if the additional five years of life are seriously compromised, with frequent pain, difficulty in breathing and the like, this also might not be the patient's sense of success. Furthermore, if only four people have ever been given this pharmaceutical, that fact is relevant to an assessment. The physician provided information but the information was partial and misleading. Polemicists and interest groups frequently exploit conceptual vagueness in making their case, sometimes intentionally, sometimes not; they take advantage of the fact that most people, prior to reflection, will assume a familiar context and will assume they know the meaning of the words being used.

As indicated earlier, there is another important aspect to 'sound' reasoning, namely, obtaining the best available evidence, in order to be as confident as possible in the claims on which the reasoning is based. Largely in the GM organism context, that will be scientific evidence. So, it is reasonable to begin with that kind of evidence. Pamela C. Ronald and Raoul W. Adamchak, in their excellent book *Tomorrow's Table* (2008), crisply articulate the challenge for public discourse and decision-making, 'So how can the public distinguish rumours from high quality science, determine what an established scientific "fact" is, and what is still unknown?' (p. 82). One thing, among many (such as including recipes), that makes Ronald and Adamchak's book unique and fascinating is that Ronald is a professor of plant pathology at the University of California Davis and Adamchak, her husband, is an organic farmer.

They cite, as an example of a rumour with absolutely no scientific research supporting it, one of the pieces of the 'evidence' provided by Jeffery Smith in his book *Seeds of Discontent* (2004). Smith has no scientific training; he was a political candidate in Iowa, USA for the Natural Law Party. Indeed, as any reader with a modicum of science background will quickly detect, he has scant scientific knowledge and a remarkably poor grasp of scientific methodology, reasoning and theorising. There is a segment of the population that will find this an attractive profile, since they view science as the root of all evil. Their alternative, however, of rumour, intuition or divine revelation has a rather poor track record – certainly a much poorer track record than contemporary scientific research. Moreover, few of the advocates of this anti-scientific stance would fail to take advantage of the fruits of science in modern medicine or in building materials or in aviation and the like, an inconsistency to which I will return.

Smith claimed that genetically engineered food fed to lab animals resulted in stomach lesions. He cites an experiment that was unable to be replicated and had several methodological and analytical flaws – all pointed out by the scientific community before Smith's book appeared. That experiment was conducted by Stanley W. B. Ewen and Arpad Pusztai (1999) and was published in *The Lancet*, a highly reputable medical journal. The findings were never replicated and criticisms of the research and the analysis of the findings were numerous (for an early exchange on this research see Lachmann, 1999). The more substantive issue, however, has to do with the choice of gene to insert in this particular GM potato. The GM potato used in Ewen and Pusztai's research had lectin genes inserted. Lectin is a known toxin; why Cambridge Agricultural Genetics (later Axis Genetics) inserted this gene in a product for human consumption is a mystery. As May (1999) pointed out, in a now famous, quip, 'if I were to mix cyanide with vermouth and I found the resulting cocktail unhealthy, I would be silly to draw the general conclusion that I should never mix drinks.' There is, of course, a lesson in the mistake of Cambridge Agricultural Genetics but to generalise from it to all GM crops is irrational.

He also provided a report on results from an 'experiment' performed by a 17-year-old, who fed genetically modified potatoes to mice. The boy claimed that the mice fed the GM potatoes behaved differently from those not fed GM potatoes. This is an uncontrolled, single-run experiment and the observations are subjective. Ronald and Adamchak (2008) make the obvious, and quite disturbing, point, 'The implication is that the public can trust this experiment carried out by a student, unhampered by scientific training but not those of the scientific community who pointed out the flaws in the original [1999] experiment' (p. 82). One might add, a scientific community whose voluminous research has led to contrary results in numerous experiments.

Of course, high-quality scientific research is not infallible, but the scientific community holds research to high methodological and analytical standards, and different research teams are constantly attempting to replicate research. Moreover, research results are constantly aligned with robust theories. Scientific theories are not, as creationists would like one to believe, simply hypotheses – speculations. Theories are integrated bodies of knowledge; they bring together, in a single framework, the sum of our current knowledge in an area of investigation. New results that are inconsistent with a robust theory require further examination. It might be that the theory has exposed a deficiency in the experiment, or it might be that some adjustment to the theory is needed. This is a dynamic process that ensures that the **whole** collection of the scientific results to date cohere – are consistent. The intuitive, rumour-based or single-instance, unreplicated, results that are so often cited in polemical works are isolated claims, and no attempt is made to develop a consistent overall description of accepted knowledge. There is an old adage from probability that is relevant here: the race is not always to the swift nor the battle to the strong but that is the way to bet. High-quality scientific research does not always result in reliable knowledge but that is the way to bet. It is puzzling that so many people are willing to bet on methods that seldom result in reliable knowledge, such as intuition, rumour, one-off experiments and the like, rather than on high-quality scientific research that has an enviable track record of producing reliable knowledge.

Returning to the important question posed by Ronald and Adamchak, 'So, how can the public distinguish rumours from high quality science, determine what an established scientific "fact" is, and what is still unknown?' (p. 82). They provide a useful set of criteria, which in summary are:

1. Examine the primary source of information.
2. Ask if the work was published in a peer-reviewed journal.
3. Check if the journal has a good reputation for scientific research.
4. Determine if there is an independent confirmation by another published study.
5. Assess whether a conflict of interest exists.
6. Assess the quality of the institution or panel.
7. Examine the reputation of the author.

Although entirely consistent with their set, I recommend focusing on four aspects of primary scientific research: the integrity of the author (principally, that potential conflicts of interest have been disclosed), the academic quality of the publication (journal or report), replication of the findings, and consistency with what is already known. **Primary** research is a direct examination (observation, experiment, model, theoretical explanation and so on) of a scientific claim. The contrast to primary research is **secondary**, or **tertiary**, research, in which an individual or group draws on primary research. This includes reviews of the literature, reports on a few pieces of primary research, and interpretations of primary research. Acceptable, credible secondary research will provide full citations of the primary research on which a scientific claim is based. If there are no references, be wary; this is a pretty good indication that you are simply being given an opinion, a rumour, or disinformation and propaganda. If there are references, some additional digging is required.

A comprehensive assessment of evidence requires an assessment of the quality of experimental design and reasoning in specific pieces of scientific research. Given the vast and growing body of research, that level of assessment is seldom feasible, even for scientific researchers not directly working on a specific problem but who work in related areas. In addition, a firm grasp of scientific methodology is required, including inductive inference. The four aspects set out here provide a robust surrogate for that more detailed assessment and allow obvious rumour and propaganda to be identified quickly and dismissed. The findings of almost all primary scientific research is reported in journal articles. The journal title can be typed into Google; reputable scientific journals as well as publishers of reports will have a website, which will give information on the journal, including its peer-review process and conflict of interest policy. Reputable scientific journals require that researchers declare potential conflicts of interest. An example of a potential conflict of interest would be research on the nutritional value of tomatoes funded by processors or marketers of tomato products. This is, obviously, only a 'potential' conflict of interest. We all have numerous *potential* conflicts of interest, some significant, some unimportant. It mostly depends on the context and effect of the conflict. In research, the important issue is whether it had any distorting effect on the research or the findings; this requires further analysis. Disclosure is the basis of most journal policies. Without disclosure, a journal editor, peer reviewers and readers may be misled. Disclosure ensures that appropriate scrutiny is undertaken to determine if there is, or is not, a conflict of interest that has biased the results. Consider as highly suspect articles in journals that do not have a peer-review process or in which authors do not have to disclose potential conflicts of interest. Peer review and disclosure of potential conflicts of interest are essential. Blogs, websites of individuals or organisations, pamphlets and the like do not meet these requirements; even when the claims made in them are by purported experts, they may be (indeed often are) no more than unreliable advocacy. If the expert, or those quoting him or her, cites the primary scientific research underlying the claims, the acceptability of the claims rests solely on that primary research. Castle (2006) and Castle and Culver (2006) offer excellent articles on citizen engagement, and expertise and authority that amplify, respectively, helpfully the point made here.

Trusting the claims of scientific experts or presumed authorities, where no primary research underwrites the claims they make, is fallacious reasoning, which logicians call the fallacy of appeal to authority. It has long been known to be a fallacy and originally was given a Latin name, *argumentum ad verecundiam*. Even though the fallacy of this type of reasoning has been known for millennia, it is still widely used and individuals, regrettably, are frequently misled. Much advertising depends on public gullibility and receptivity to this fallacious reasoning. Without doubt, scientific experts are better placed to steer others to the primary research and provide some guidance on how to interpret it, but the foundational evidence remains the primary research; scientific claims by experts without the support of primary research is opinion and no more reliable than any other opinion. Of course, in legal contexts, expert testimony is common – even on scientific matters. Three factors are important in this reliance on expert testimony. First, in most cases, the matters at hand are not amenable to scientific research, even though scientific research might be relevant. Whether this person, on this occasion, in the manner claimed is guilty of assault is not amenable to direct scientific research. There is usually no shortage of evidence brought forward – some scientific but much that is not – but the veracity, or not, of the charge of assault is not amenable to scientific resolution. Second, when an expert gives scientific testimony, there is a presumption that she or he can provide the relevant primary research; it is, after all, regarded as testimony, not proof. Third, and connected to the second, experts are examined, cross-examined and re-examined as part of the process. Both sides can produce experts, both sides can challenge the experts' claims, both sides can request the primary research, and both sides can make clear to a judge or jury that the testimony is 'the opinion' of the expert and let the judge or jury sort out how much weight to give that opinion. Where an expert's claims go unchallenged, one can assume both sides accept that there is primary research to support the claim.

A sense of the quality of a journal – its ranking in the field – also can be gained by a Google search such as, 'ecology journals ranking'. This is somewhat crude but does dig one level deeper into the reliability of evidence. Ultimately, to do a thorough investigation of the evidence, one has to obtain the article and read it. This will be difficult for those who do not have access to an academic library, although most journal articles are now online and can be obtained from the publisher for a fee. In addition, without some training in a field, making sense of the data and commentary will be daunting. Fortunately, a great many of the claims that are mere opinion, rumour or disinformation will be exposed after preliminary investigation. Many will have no references to primary research, and many more will have references to obviously suspect sources or to non-primary research sources, which on investigation themselves lack references to primary research.

Scientific evidence is the most relevant evidence when claims of environmental, health, or yield benefits or harms are being made, but other kinds of evidence will be more relevant to claims about interfering with life or the benefits and harms of patenting life. Moreover, even if some consensus can be reached on harms and benefits, balancing harms and benefits (making trade-offs) will involve values. In these cases, sometimes people will proffer theological/religious evidence or principles; sometime ethical principles will be pressed into service. These non-scientific (non-empirical) domains of discourse and decision-making involve quite different kinds of evidence and methodology, which are nonetheless important. The next three sections of this chapter tackle some of those. Suffice it to say here that the logical requirements of reasoning set out above apply equally in these domains as in science; their differences from the domain of science, and from each other, lie in the justification for accepting fundamental claims.

To this point, the components of sound reasoning examined have been consistency, rules of inference (deducing a claim from other claims), clarity of concepts and justified evidence. There is another tool worthy of mention, although it is not a rule of inference, and is not deductive. Some philosophers and logicians have regarded it as an instance of induction. This is a matter of much controversy. For our purposes, we do not need to elevate it to a rule of inductive inference and can avoid the controversy; it is sufficient to consider it to be a heuristic tool of reasoning. That tool is analogy. In biology, the best-known use of analogy is found in Charles Darwin's *On the Origin of Species*. Darwin drew an analogy between artificial selection and natural selection. Natural selection, in Darwin's time, was not observable; today we have a wealth of observational evidence (see Ridley, 1996, which provides a detailed discussion of natural selection and cites extensive primary research). For Darwin, and others at the time, the effects of natural selection were clearly observable; the mechanism (process) was not. What Darwin needed was a way to connect what he believed were the observable effects (biogeographic distribution of organisms, lineages of changing anatomical features, geological patterns of fossils and so on) to what he believed, correctly, was the cause of those effects, namely natural selection.

Animal breeders from pigeon fanciers to racehorse breeders to agricultural animal breeders knew that selecting animals with desirable traits allowed the breeder over time to increase the number of animals with those traits. This is artificial selection because it is not a process of nature but of human 'selection'. That artificial selection is successful establishes that selection leads to changes in organisms. The analogical leap is to claim that the artificial process has a natural analogue. The invisible hand of nature selects with the same results as human selection – changes in organisms. The invisible selecting hand was, for Darwin, a function of the struggle for survival. More organisms in a species are born than the natural resources can support. Some traits make organisms better at surviving (more efficient nutrient extraction from food, speed, size, cognitive skills and the like). Those with advantageous (survival enhancing) traits will, on average, have more offspring (known today as differential reproduction success). After many generations those with advantageous traits will vastly outnumber those that do not. Nature, consequently, has selected some traits over others. This natural process has no intentions involved; it is as mechanistic as the water cycle and the rising and receding of the tides. This is a brilliant use of analogy. As Michael Ruse (1975, 1999) has pointed out, Darwin's use of analogy can be traced to the philosophy of science of William Whewell (1840). It provided Darwin with a way of **revealing** a hidden cause by exploiting its analogue, artificial selection, which was a known and demonstrable cause. Not surprisingly, given the passage of 150 years, evolutionary biologists today have a more nuanced view of the invisible selecting hand of nature; for example, many more things lead to differential reproductive success than the struggle of organisms against nature. For those interested in exploring the richness of contemporary evolutionary biology, Mark Ridley's (1996) book is an excellent place to start.

Another famous use of analogy is William Paley's watchmaker argument for the existence of God. Paley, in *Natural Theology* (1802), argued that the universe, like a watch, manifests intricate design and indeed we know it was designed. No reasonable person would believe that a watch just fell together without a watchmaker governing a design and assembly process. Since the universe manifests the same design and precision of its parts and their interdependence, there must be a universe maker – God. The schema is:

Watch = Universe
Watchmaker Universe maker

Darwin's use demonstrated the power of analogy, especially its heuristic power. By contrast, Paley's use of analogy revealed important potential deficiencies in the use of an analogy, as was pointed out by David Hume and others. A major potential deficiency is that analogies always have embedded disanalogies. If a comparison of two things, processes and the like have no disanalogies, they are equivalent and hence not just analogous. Paley had in mind the Christian God, the universe creator whose creative deeds are outlined in the book of Genesis in the Bible. The analogy does not come close to proving that this Christian God exists, nor does it reveal important features of such a God. A watchmaker is fallible, limited in knowledge, limited in power and not wholly good. Hence, by analogy, God is fallible, limited in knowledge, limited in power and not wholly good. We know that these are disanalogies because the Bible tells us the true properties of God. But, if you have this independent source of knowledge, what work is the analogy doing? No work; indeed, in this case, it is actually revealing inaccurate knowledge about God. The real source of knowledge, and the correction of the analogically derived properties, is the Bible. It, however, is only a source of knowledge if you already accept that God exists and that the Bible reveals his properties and deeds. In this case, the analogy adds nothing to one's knowledge.

There is another lesson to be learned from this analogy. Just as there are disanalogies between the watchmaker and the Christian God, there are disanalogies between a watch and the universe. The universe is dynamic; it changes over time. Stars are born and decay, solar systems disintegrate and so on. This led David Hume (1739, 1777, 1779), although writing earlier than the publication of Paley's *Natural Theology*, to suggest that maybe a better analogy would be an animal; Hume also provided a wealth of counter-examples to the Paley-type analogy. The animal analogue has fewer disanalogies than the watch; animals are conceived, born and constantly change. If the universe is more analogous to an animal, then, since there is no animal maker, there is no universe maker. A Christian might be tempted to respond that the animal does have a maker, namely God. But that simply builds into the analogy the very conclusion it is supposed to be proving:

Animal	=	Universe
Animal maker (God)		Universe maker (God)

The conclusion of this analogy is built into the analogy; the maker in both cases is God. In logic, this fallacy, which Aristotle sets out, is known as *petitio principii* (begging the question). Today, it is common for people to use the phrase, 'that begs the question', to means something like 'that calls out for the question to asked'. The suggestion is that some comment or argument triggers another question, which must now be answered. That is not the original meaning of 'begging the question' or 'begs the question'. To avoid confusion, most philosophers, including logicians, now refer to the fallacy as a circular argument or circular reasoning.

So, as Darwin showed, analogies can be a powerful tool of reasoning – albeit a heuristic one – but, as Paley's analogy shows, the tool is fraught with difficulties and great care must be taken. The two lessons from Paley's analogy are: (1) ensure that the two principal analogues are as analogous (similar) as possible and as needed for the purpose at hand, and (2) ensure that the claims one wishes to support by using the analogy can actually be drawn from it. Paley's analogy failed on both counts. Analogies abound in science and in all other areas of reasoning, so knowing how to use them and assess them is crucial. In the next section, I provide an example, which uses the Frankenfood label, of the use and abuse of analogy in the GM context.

3.2 Relevant ethical theories

In informal colloquial discourse, 'theory', 'hypothesis' and 'conjecture' are used more or less interchangeably; this is not the case in ethics and in the sciences. In colloquial discourse, the expression 'theoretically speaking' conveys uncertainty. This uncertainty and tentativeness in ethics and sciences is usually expressed as 'hypothetically speaking', where a hypothesis is something closer to a guess, or a conjecture, that is yet to be tested or established. In ethics, as in the sciences, 'theory' is a technical term; 'theoretically speaking' means that a claim is 'in accordance with a widely accepted theory'; that is, it can be proved by the theory (ideally, can be deduced from it). Theories in ethics and the sciences are similar in an important respect; in both, theories are the foundation and underpinning of knowledge claims. They differ mostly in the justification provided for accepting them and the nature of the knowledge claims. Any attempt to align colloquial discourse with the scientific and ethical meaning of theory is almost certainly doomed to failure, but I trust that for the remainder of this book I will be understood to be using theory in its technical sense.

In the sciences, a theory is a comprehensive, well-confirmed, interconnected body of knowledge. Einstein's general theory of relativity, quantum theory, and the modern synthetic theory of evolution are just such interconnected bodies of knowledge. If something is deducible from a theory, *it has the highest degree of certainty possible*. To doubt a claim that can be deduced from a theory is to doubt the entire edifice of knowledge in that domain of science. Typically, a domain of science embraces only one theory at a time – the theory that best accounts for the phenomena in that domain. Of course, newly discovered phenomena (observations or outcomes of experiments – manipulations carefully designed to uncover aspects of nature) may require some revisions to a theory (adjustments to a generally accepted framework). In rare cases revisions may be insufficient to allow new phenomena to be encompassed by the current theory. In such cases, the development of a new theory is required (a complete reorganisation of the interconnections of our knowledge claims or, more radically, a change in what we accept as knowledge claims). After some period, the new theory will replace the old one. This is not a frequent occurrence, so examples are few; Copernicus–Galileo replacing Ptolemaic astronomy, Galileo–Newton replacing Aristotelian mechanics, Einsteinian relativity replacing Newtonian mechanics, and Darwinian evolution replacing a teleological, and species immutability, framework are frequently cited examples. A few more cases could be added but not many.

The touchstone for scientific theories is the entire collection of phenomena as experienced within a domain such as physics, chemistry or biology. Theories are used to integrate our claims about phenomena and to explain and predict them. When a significant collection of phenomena cease to be explainable by the theory or the predictions made by the theory fail in worrying ways (e.g. many inaccurate predictions or a few, but critical, inaccurate predictions), the theory needs attention. It needs to be either revised or replaced. Consequently, scientific theories rise or fall by their ability to integrate, explain and predict accurately all the phenomena in that domain.

Ethical theories differ both in the touchstone for acceptance and the kinds of claims that follow from the theory. Scientific claims are claims about what exists in the world and how things behave; they are claims about things. Ethical claims are value claims – claims about what is good, desirable and acceptable (or bad, undesirable and unacceptable), claims about what ought, or ought not, to be done in particular circumstances. The touchstone for ethical theories fundamentally comes down to what goals one wishes to achieve. Goals are, of course, themselves values. Generating consistent judgements about behaviours that result in a viable society would be an example of such a goal. A society in which some individuals had so little to lose that they murdered, stole from or in some other way negatively affected others is not viable (it will degenerate and cease to be a functioning 'society'), and any ethical theory that generated value judgements that justified such a social structure would be rejected.

Different goals will justify different ethical theories, and this is, in part, why different individuals and groups adopt different ethical theories. Three things about this plurality of theories need to be underscored. First, the divergence of opinion on goals is not wide. Second, contingent facts about the world are important. For example, it may be that the available resources in a society are insufficient to meet the basic needs of the individuals in it. An ethical theory that generates principles that suggest that individual needs should be met cannot be faulted for any social breakdown arising from the contingent resource factors. The fundamental principles embedded in the theory do not justify behaviours responsible for any social breakdown; it is the current circumstances that frustrate the realisation of those principles. Such a theory may be deemed not useful in the circumstances (idealism is often the epitaph used to describe such theories), but it is not the theory as such that is flawed; it is its applicability under the circumstances that might be questioned. Sometimes an inapplicable theory needs to be replaced with one more suited to the circumstances. Sometimes, however, focusing on the ideal identifies the contingent barriers and motivates individuals and groups to work to change those circumstances. Aiming high, the reformers in Victorian London slowly transformed the plight of the poor and changed the social fabric for the better. That science and technology played an important part in achieving these changes is worth keeping in mind.

Third, not just any value-generating framework will be justifiable. Obviously, ethical theories that generate value judgements that frustrate achieving the accepted goals will not do, but there are other, goal-independent criteria as well, criteria that apply equally to scientific theories and that were set out in Section 3.1 of this chapter, such as consistency and clarity. Hence, the ethical theories set out here assume that ethical discourse is rational. I note, in passing, that there is a school of ethics that views ethics as non-cognitive. For this school, ethical statements do not state facts and are neither true nor false. Hence moral reasoning, to the extent it is possible at all, does not adhere to the criteria of reasoning set out above. This is a robust position but far from mainstream and not the one adopted here.

Ethical theories in Western nations divide into two general foundational approaches: teleological theories and deontological theories (Rawls, 1971). The word 'teleological' derives from the Greek τελοσ (*telos*), meaning 'end and completion', and λογοσ (*logos*), meaning 'word, truth and study': hence, biology, the study of life (Liddell, 1966). 'Deontological' also has its roots in Greek; δεον (*deon*) means 'that which is binding (obligation), needful, right and proper' (Liddell and Scott, 1966). Often teleological theories are referred to as 'consequentialist' theories to emphasise the central role that the consequences (ends) of an action play in determining the moral status of the action. Egoism and utilitarianism are two well-known teleological theories. For both, the ethical assessment of actions is based on consequences. In the case of egoism, what matters are the consequences for a specific individual, namely, the individual doing the assessment. At the heart of sophisticated versions of egoism is enlightened self-interest. In the case of utilitarianism, what matters is the consequences for the whole, hence the maxim, 'the greatest good for the greatest number'. Assessing actions using some version of utilitarianism is familiar, though not necessarily acceptable, to everyone.

By contrast, deontological theories involve duties and obligations that apply, for the most part, regardless of the consequences. The qualifier 'for the most part' is important; properly understanding how to apply a deontological principle, such as 'do no harm' requires knowledge of the consequences of a behaviour. So, obviously, it is not that consequences are irrelevant in deontological theories; it is that they are subservient to principles. For consequentialist theories, consequences (outcomes) generate the principles and, therefore, have primacy. An oft-used example of the importance, and place, of consequences in deontolological theories (in effect, the importance of context) is a surgeon amputating a limb (thereby doing harm in the vernacular sense of harm) to save the person's life. This *prima facia* harm in this context turns out not to be a harm at all but a benefit to the individual. This illustrates how complex something as seemingly straightforward as harm can be even for someone who accepts the principle, 'do no harm'. Kagan (1998) provides a clear exposition of the various deontological positions on harm. When we examine the precautionary principle and later the concern that biotechnology is an unnatural interfering with life that as a matter of principle is wrong, the complexity will increase.

In this section, four ethical theories are examined: natural law, Kantian ethics, social contract and utilitarianism. I also provide a sketch of ethical naturalism or, rather, a version that I think is defensible. Ethical naturalism is less an ethical theory than a stance with respect to the foundation of ethics. These are the ones most frequently pressed into service in debates over agricultural biotechnology. Natural law theory (a deontological theory) underpins many positions on risk aversion (as in one version of the precautionary principle; see Section 3.4) and the position that biotechnology is manipulating the essence of life (interfering with natural order) and on principle is wrong (Section 4.2). In addition, with different nuances for different religious communities, it lies behind many theologically based ethical theories. Hence, I spend a considerable amount of time explicating it. Ultimately, as I signalled earlier, the moral judgements about actions in biotechnology converge whether one adopts a natural law position or a utilitarian position. The ethical foundations are different and the specific steps in the reasoning processes are different but the moral judgements about biotechnology converge. What makes them converge is the sensitivity to context in applying natural law principles and the need to prioritise them when two or more justify inconsistent moral judgements. This convergence is exposed in Section 4.2.

Kantian ethics (also a deontological theory) has had a profound influence on ethics and applied ethics in Anglo-European countries (European and English-speaking countries such as the UK, Canada, the USA, Australia and New Zealand). Accepting the foundational principles of Kantian ethics is usually what motivates concerns about people being treated solely as means to an end or about unequal distributions of benefits and harms. As such, elements of this theory make their way into disputes about capitalism, globalisation and multinational corporate control, where some see these as reducing individuals to means (instruments: mere pawns) in the production of corporate profits that benefit the few at the expense of the many (as discussed in Section 6.1 and Chapter 8). The essential elements of Kantian ethics are embedded in social contract theory. Hence, I concentrate mostly on it and simply set out the central Kantian principles.

Social contract theory has special relevance to balancing rights, responsibilities and the satisfaction of interests in a viable and just society; something that is obviously relevant to balancing harms and benefits in agricultural biotechnology. One of its foundational concepts is that rational individuals pursuing their own self-interest can form social groups in which the self-interest of each individual, though constrained, is maximised. Going it alone is assumed not to be in an individual's rational self-interest; tacitly accepting the rules of a social collective that maximise the satisfaction of everyone's interests, in accepting minimal constraints, will lead to more of one's interests being satisfied than going it alone.

Utilitarianism is attractive to many because it advocates the maximisation of the satisfaction of the interests of the greatest number of individuals. In addition, it seems ideally suited to risk analysis. As we will see in Section 3.3, risk analysis involves identifying benefits and harms, balancing them, and mitigating or managing harms worth risking to achieve the potential benefits. This seems very consequentialist; harms and benefits, on the surface, appear to be mere consequences. Risk analysis involves maximising benefits and minimising harms, which seems akin to the utilitarian maxim of seeking the greatest good for the greatest number. This superficial similarity, however, is deceptive; risk analysis is compatible with other theories. Almost all versions of natural law have a principle something like, do no harm or do as little harm as possible under the circumstances. It is the empirical assessment of harm and the circumstances that make risk assessment as relevant to moral judgements based on natural law as to those based on utilitarianism. Similarly, social contract theory seeks to maximise the satisfaction of individual interests consistent with functioning society, entailing that some individual interests cannot be realised. Hence, although utilitarianism and natural law are often portrayed as antithetical – and in many ways they are – they have many points of convergence on moral assessments of biotechnology. This is largely because balancing harms and benefits is essential in both frameworks and, as subsequent chapters will demonstrate, lies at the heart of the controversies over biotechnology; hence, both frameworks require some technique of risk assessment. This convergence notwithstanding, it is important for anyone venturing into an examination of the ethical, social, public policy, and legal and regulatory aspect of biotechnology to be clear on how these two frameworks differ and, consequently, to appreciate more clearly why in this case they converge. It is this that justifies the attention paid to the four theories in this section.

The main reason for including ethical naturalism in this section is that it is a view I adopt and, hence, knowing some of its features ensures that my appeal to it, often tacit, can be identified by a reader. In a book of this kind, where ethical, epistemological and metaphysical commitments lie at the heart of positions and arguments, a reader should not have to ferret out the theoretical and conceptual commitments of an author. Consequently, to be transparent about these, I declare my commitment to naturalism and make explicit that the ethical theory that I combine with my naturalist stance is social contract theory. Now to a fuller exposition of each.

3.2.1 Natural law theory

Natural law theory has a very long history stretching back to Plato and Aristotle; Aristotle's distinction between conventional justice and natural justice is rooted in this theory. Although it has undergone many revisions and adjustments over the last two millennia, it has persisted into the present. In essence, as the name suggests, natural law connects what is right and wrong conduct to what is natural.

Cicero, in *De Re Publica* (51 BCE – Before the Common Era[1]) sets it out this way (emphasis mine):

> True law is **right reason** in agreement with **nature**: it is of universal application, unchanging and everlasting... Whoever is disobedient is fleeing from himself and denying his human nature, and by reason of this very fact he will suffer the worst penalties, even if he escapes what is commonly considered punishment. (Buckle, 1993, p. 164)

About 1,300 years later Thomas Aquinas in *Summa Theologica* (1265–1273) captures its essence this way (again, emphasis is mine):

> Whatever is contrary to the **order of reason** is contrary to the **nature of human beings** as such; and what is reasonable is in accordance with human nature as such. The good of the human being is being in accord with reason, and human evil is being outside the order of reasonableness... So human virtue, which makes good both the human person and his works, is in accordance with human nature just in so far as it is in accordance with reason; and vice is contrary to human nature just in so far as it is contrary to the order of reasonableness. (Buckle, 1993, p. 165)

The important elements in both are as follows:

1. Reason is inextricably connected to human nature, right action and virtue.
2. Failure to live according to the dictates of right reason debases the person (corrupts that person's nature).

[1] In an attempt to secularise European dating based on Christianity but also retain the dating scheme used for centuries, scholars have replaced before Christ (BC) and Anno Domini (AD: in the year of our Lord) with before the Common Era (BCE) and the Common Era (CE). This is a bit artificial and cumbersome but is widely used, so I have adopted it here. Unless clarity requires it, unmodified dates are CE; hence, 2009 is understood to be 2009 CE.

3. The goodness of the human person (virtuousness by pursuing right reason) and the goodness of the actions (works) of the person are inextricably connected.
4. True law (the consequence of right reason in agreement with nature) applies universally, and is unchanging and everlasting (Cicero states this explicitly in the quoted passage; Aquinas holds this as well); in today's language ethical principles apply to everyone, in all places and in all times.

These millennia-old philosophical phrases and constructions obscure for most people the intellectual depth and current attractiveness of this theory. Before examining its twenty-first-century version, however, it is worth gaining a bit of clarity on the earlier view since it still underpins the modern version. This Cicero–Aquinas version, unlike most other ethical theories currently employed, considers right action to be a natural, indeed inevitable, product of the goodness of the person. A corrupt person, by definition, cannot perform a good action. It might be the same action that a good person would perform under the same circumstances. The corrupt person, however, performs it for the wrong reasons and by virtue of that it is a tainted action. This takes us to the heart of the view; the reasoning person is a good person. Aristotle defined human beings as 'rational animals'. Cicero and Aquinas are connecting that to ethics; to fail to be rational is to fail to be human; to fail to be human is to fail to behave ethically. Animals may perform actions that we admire, and for which we reward them, and they may perform actions that we dislike, and for which we punish them; we may breed them or train them to maximise the behaviours we admire or from which we benefit. Nonetheless, their behaviour is not ethically good or bad; only rational animals (humans) can be ethical, and that is because, *alla* Aristotle, they are rational in addition to being animals. Hence, to abandon one's rationality (to be driven by crass desires and irrational urges) is to cease to be fully human – to become, in the extreme, a mere animal.

Common contemporary aphorisms expressing these ideas abound; for example, there is no virtue in doing the right thing for the wrong reasons, he is no more than an animal, or he is behaving in accord with his animal self.

It is easy to see how, on this version of natural law theory, if a person has murdered someone, executing him is acceptable; he has become an animal and lost any status as a rational animal. Indeed, execution may be required because, having abandoned his humanity in this most dramatic of ways, he requires execution to avoid further carnage, in the same way as an animal that has attacked a human is killed. This is not a deterrence argument. This is an ethical necessity. Contemporary versions have had to wrestle with the development over the last century of the concept of rehabilitation. If a person can be rehabilitated – brought back into humanness – do others have an obligation to attempt to do this rather than execute him? More about this later.

There is, of course, an immediately obvious problem with the reliance on reason. People of goodwill, thinking carefully and rationally, can come to different conclusions about important and complicated matters. Just whose reason is the right reason to which Cicero alluded? It is not adequate to claim that murder – or this or that behaviour – is wrong, and therefore those engaging in it have lost their humanity and their rationality. The theory demands that rationality (right reasoning) determined the goodness of the person and her actions, not the goodness of the actions determining right reasoning. To go both ways is circular reasoning, an instance of *petitio principii* (and a tight circle like this is clearly bad reasoning).

There is also the challenge of the last century of scientific findings. The foremost one is that the dividing line between animals with cognitive capacities and humans is virtually non-existent, and attributing rationality to humans is fraught with problems; most people, at least some of the time are irrational – a feature advertising gurus, and many others, exploit shamelessly. Moreover, some humans are clearly mentally challenged; for them rationality is elusive.

There is yet a third, and important challenge, to the internal logic of the theory. This challenge has two related faces: Hume's 'is-ought' barrier and G. E. Moore's naturalistic fallacy. David Hume (1711–1776) pointed out that any reasoning process which results in a value statement (right action, what one ought to do and so on) will have among the premises on which it draws (i.e. its evidential basis) a value claim (Hume, 1739).

To claim, 'One ought not to engage in war' requires justification. A justification appealing to the nature of things might be, 'War results in death, destruction, human suffering and the loss of loved ones.' This may be a correct factual claim but, as Hume pointed out, it alone cannot rationally justify the value claim that 'One ought not to engage in war'. An additional premise is required, something like: 'Activities that result in death, destruction, human suffering and the loss of loved ones ought not to be engaged in.'

But this claim is a value claim and itself requires justification, and so on. An argument that has a missing, suppressed or 'taken-for-granted' premise is known as an enthymeme. An argument that appears to derive a value conclusion solely from factual premises is, according to Hume, an enthymeme; there is at least one value premise missing, suppressed or taken for granted.

Natural law appears to generate value claims from only factual claims – the nature of things, and hence involves fallacious reasoning.

To underscore the importance of Hume's point and to connect this back to the discussion of analogy in the previous section, consider Greenpeace's use of the term 'Frankenfood' to describe GM food. The force of this label rests on an analogy almost everyone will immediately construct. It is of course an elision of 'Frankenstein' with 'food' – Franken(stein)food. Few, in rich countries, have not heard of Mary Shelley's novel *Frankenstein* (1818), in which Frankenstein creates, from collected body parts, a monster. So the surface analogy is something like:

$$\frac{\text{Frankenstein constructed a creature}}{\text{A destructive monster is unleashed}} = \frac{\text{Scientists have constructed GM food}}{\text{A destructive food is unleashed}}$$

This analogy may be evocative, and therefore useful for propaganda purposes, but there are several important features that weaken its argumentative force. First, and obviously, *Frankenstein* is a novel; no such creature has been created and, hence, any description of an outcome is speculative. Consequently, any inference from Shelley's chilling speculation about a possible outcome for Frankenstein's monster on the outcome of an actual manipulation of plants is entirely speculative. Second, and crucially important, the two morals one is supposed to draw from the analogy are that GM biotechnologists, like Frankenstein, are doing something unnatural (meddling with nature), and that meddling with the natural order is **wrong**. This is the value premise that, as Hume taught us, has been suppressed. Now uncovered, it requires some examination and justification. Why should someone accept that meddling with nature is wrong? There are two common tacks to answering this. First, there are people who hold theological views that they claim entail this value. An immediate problem for this tack is that we meddle with nature all the time in medicine and few would want their theological views to halt medical advances. There are other features of this tack that I explore more fully in Section 4.2.

Second, advocates of this value claim hold that meddling with nature is inherently dangerous and, hence, **wrong**. This requires that the claim that meddling in nature is inherently dangerous be substantiated. This is an empirical matter. The material in Chapters 5–7 examines aspects of this claim. Suppose, however, one grants this empirical claim. The more significant problem with this tack is that it rests on yet another value claim: actions that are inherently dangerous are wrong. However, heli-skiing, white-water

Table 3.1 *Traffic fatalities and injuries by region (created by R. Paul Thompson from public domain data of the World Health Organization)*

Region of the world	Traffic fatalities	Traffic injuries
Africa	170,118	6,116,559
Central and South America	125,959	4,410,736
China	178,894	5,384,909
Eastern Mediterranean	71,600	2,563,750
Europe	172,856	5,295,425
India	216,859	7,203,864
South-east Asia	118,608	3,997,631
Western Pacific	66,495	2,205,377
North America	49,304	1,670,374
Total	1,170,693	38,848,625

rafting, mountain climbing and the like are also inherently dangerous. Many may consider that engaging in these activities is imprudent but only a small minority would deem them immoral (morally wrong). Furthermore, driving an automobile is inherently dangerous – to the driver but also to others, even pedestrians – as the World Health Organization Data for 1998 show (Table 3.1).

A vanishingly small number of people, however, would deem automobile use morally wrong. So the underlying moral value is highly suspect, undermining the entire argument. As will be seen in the next section, the most appropriate response to activities that are inherently dangerous is risk analysis and risk management – not a leap to moral disapprobation. No doubt, if meddling in nature were always catastrophically dangerous with almost no benefit, moral disapprobation might seem appropriate (a claim some anti-GM advocates seem to endorse). But, as we shall see in later chapters, there is no evidence to suggest that this claim is remotely credible. Like most human actions, this one involves the risk of harm but also confers benefits; balancing and managing these is the rational course of action. This example highlights the value of applying analogical reasoning and ethical theorising to the GM food controversy.

G. E. Moore, an early twentieth-century philosopher (1873–1958), argued that ethical attributes such as 'good' are non-natural attributes – where I use the less philosophically complex term 'attribute', Moore actually used 'properties', which are ascribed to things and actions (Moore, 1903). An example of a natural attribute of an object would be its colour or its texture. A natural attribute of an action would be the time and place of its occurrence or the sequence of events involved. These are part of the nature of things and actions, and hence natural attributes. Unlike these, the attribute good (morally good) when attributed to an object is non-natural; it cannot be found in nature. Moore justifies this position by what he called an open-question argument. Imagine any natural attribute of a thing – the brilliance of its colour, for example. In every case of the attribution of a natural attribute, the question can be posed, 'But is it good to possess that attribute?' Consequently, 'good' cannot be a natural property, since the attribute good already answers the open question whereas a natural attribute does not. To mistakenly conclude 'goodness' from the natural attributes of a thing or action is, according to Moore, to commit the naturalistic fallacy – a fallacy of reasoning. It is obvious that this is a frontal assault on the version of natural law theory set out so far. If moral attributes are non-natural, then they cannot be derived from the nature of things, and, worse, to do so involves fallacious reasoning. For a theory grounded in 'right reason', accusing natural law theorists of having a fallacy of reasoning embedded in the theory strikes at the core of the theory.

A potential solution to many or all of these challenges and one adopted by many natural law theorists is to appeal to truths that no rational person could deny: self-evident truths such as the sanctity of life, protection of the dignity of the person and equality of treatment. These are value claims but they are ones no rational, thinking person would deny; they arise from *rational reflection* on the nature of things and underpin other truths generated by reason. They are not derived from the nature of things (so there is no fallacy of reasoning) but are intuited from the nature of things by rational reflection.

This has been, and indeed still is, an attractive revision. The US Declaration of Independence uses this language; its second paragraph states:

> We hold these truths to be self-evident, that all men are created equal, that they are endowed by their Creator with certain unalienable Rights, that among these are Life, Liberty and the pursuit of Happiness.

Observe that the word 'rights' is used here. Although the concept of a right goes back to at least Roman times, the concept of rights played, at best, a minor role in natural law theory until the sixteenth century. It is worth noting also that the US Declaration explicitly connects self-evident truths and unalienable rights. The self-evident truths are that 'all men are created equal' and that they 'are endowed by their Creator with certain inalienable rights'. It then gives three examples of these rights – rights that arise from a self-evident truth.

This use of rights in a quasi-natural law context owes a lot to the writings of John Locke (1690, 1691) and his contemporaries. Locke's views are complex; although he clearly used the language and many of the ideas of natural law theory, he also espoused elements of what is now called social contract theory. I consider Locke to be a transitional theorist with one foot in natural law theory and another in social contract theory. Reason enables the determination of whether an institution is legitimate and its assumed powers and functions are legitimate. Reason permits an assessment of whether the institution serves to maximise the welfare, including liberty and just treatment, of its citizens. Reason and not authority (whether ecclesiastical or state) or superstition should be the basis for arriving at truths. Reasoning from evidence establishes empirical truths. Accepting and reasoning from self-evident truths establishes social, political and ethical truths.

Whether Jefferson, who drafted the US Declaration of Independence, was modelling it on Locke is not a settled issue (Wills, 1978). Locke's tripartite exposition of inalienable rights is 'life, liberty and the protection of property'. That Jefferson used 'pursuit of happiness' in place of 'protection of property' might suggest a departure from a central tenet of Locke's political philosophy. On the other hand, it might be that 'pursuit of happiness' – an expression connected, at the time, to Enlightment views on measurable social and political goods – actually encompasses 'protection of property' since it will be among the elements of happiness and a measurable element of it.

The United Nations (UN) Universal Declaration of Human Rights (motivated and influenced by the atrocities of the Second World War) has another employment of the language of rights; it makes no reference to self-evident truths. It begins:

> Whereas recognition of the inherent dignity and of the equal and inalienable rights of all members of the human family is the foundation of freedom, justice and peace in the world . . .

In this statement, the grounding of the inalienable rights is not in self-evident truths but in the requirements for 'freedom, justice and peace in the world'. This formulation could arise from at least two other theories we will examine – theories that do not rest on self-evident truths – social contract theory and utilitarianism.

If there are truths that no rational person can deny (self-evident truths), then natural law theory might circumvent the Hume–Moore challenge but neither Hume nor Moore will find this revised version of natural law successful. One thing that self-evident truths do provide is a non-arbitrary criterion of 'right reason'; 'right reasoning' is reasoning from self-evident truths. It is unlikely, however, that Hume or Moore will find this revised version of natural law successful. For Hume, reason is incapable of motivating moral action; motivation to moral action requires passion – quite the opposite of reason – so reason cannot be the basis of morality. Hume, in effect, rejects the very foundation of natural law. To the extent that self-evident truths are natural, Moore rejects them as ethical truths, since ethical truths are non-natural. If, however, the self-evident truths of natural law are non-natural, in what meaningful sense can the theory be said to be a **natural** law theory – one based on the nature of things?

Moreover, the viability of this version of natural law requires self-evident truths – truths no one can rationally (reasonably) reject. In a single culture, such as European Christendom, there is a chance that everyone will find some truths self-evident. If a community shares a common religion, history of ideas, socio-political framework and traditions, a number of ethical-political ideas may appear to be self-evident to the members of the community; the shared religion, tradition, etc., will have embedded in them values, and members of the community will have those values inculcated into them. Hence, it is not a surprise that those values will appear self-evident. But, on a larger temporal, geographical, cultural and religious canvas, this seems untenable. The world's tapestry of diverse ethical communities, cultures, and religions guarantees a plurality of self-evident truths. It is noteworthy that the UN Declaration, unlike the US Declaration, wisely, given the pluralistic, culturally diverse nature of the world, does not proclaim its inalienable right as self-evident truths but rather as requirements to achieving ends that it hopes all peoples and nations can be convinced are worth achieving: freedom, justice and peace.

3.2.2 Kantian ethical theory

One of the most influential deontological theories derives from the work of Immanuel Kant, the eighteenth-century Prussian philosopher (1724–1804).

His view has exerted so much influence on current formulations of deontological theories that they are increasingly being called, or at least identified with, 'Kantian ethics' (Beauchamp and Childress, 2009). Kant is credited with solidifying the place of respect for others in Western ethics. His view is often encapsulated in the phrase, 'Treat others as ends-in-themselves and not as means.' Kant did not put this later formulation of his categorical imperative in quite this way. Kant (1785) asserted, 'Act always so that you treat humanity, in your own person or another, never merely as means but also at the same time as an end in itself.' It is worth noting that Kant did not proscribe treating someone as a means, contrary to what the later, shorter, dictum might suggest, but required that one must simultaneously be treated as an end. This idea has had a profound effect on ethics although its exact import has been the subject of much debate. What seems commonly accepted is that all the formulations of Kant's categorical imperative are inextricably tied to personal autonomy (a connection he likely adopted from Rousseau's *Social Contract* of 1762). Kant wrote in 1760 that it was Rousseau who taught him the value of humanity (Korsgaard, 1992). To treat, or regard, others as ends-in-themselves is to accept that they are autonomous agents capable of rational deliberation and of making choices. H. J. Paton (1971) provides an excellent discussion of the categorical imperative and autonomy.

The challenge in building Kant's insight into an ethical theory is to find a way of judging when a person is being treated as a 'mere means' or as an end. Put another way, how does one know when one is respecting the autonomy of another? The only lasting and compelling answer to this relies heavily on the assumption that the way one understands respect for one's own autonomy can be generalised to others – something akin to 'Do unto others as you would have them do unto you.' Kant's first formulation of the categorical imperative, in Section 2 of the *Groundwork for the Metaphysics of Morals*, states, 'Act only in accordance with that maxim through which you can at the same time will that it become a universal law' (Kant, 2002, p. 37). This formulation has been captured in the concept of 'universalisability' (Hare, 1952, 1964). Kant appears to have regarded these two formulations of the categorical imperative as equivalent. The principle of 'universalisability' has affinities with the generalisation from one's own case to that of others, although it is more properly the inverse; if you cannot will that others do it, don't do it yourself.

There are well-known problems with generalising from one's own case. Here I sketch two. First, in order to apply the results of reasoning in accordance with the principle of universalisability, one virtually always needs to employ a *ceteris parabus* clause (Latin, meaning 'other things being equal'; only when circumstances are the same in relevant respects do the same properties apply to them). How one wishes to be treated depends a great deal on the particular configuration of circumstances. Since others will never be in exactly the same circumstances – especially since some of those are internal, such as past experience and mental framework – some adjustment must be made to the results of reflection on one's own case. Even something as dramatic as taking the life of another is not without problems. For example, if you are a healthy 30-year-old, you might well conclude that you are prepared to act on the maxim, 'do not take a human life except in self-defence', and you are prepared to generalise it. But, if you are an ailing 80-year-old or someone in the later stages of a neurodegenerative disease, you might have a different view of the acceptability of that maxim to your case. It is worth noting, that the terms 'killing' and 'murder' are often used interchangeably but this is unwise; 'murder' means morally wrong killing, whereas 'killing' is descriptive. Whether killing in self-defence, in war or in retribution for capital crimes is murder depends on whether that kind of killing is wrong, an issue fraught with controversy.

The second problem centres on how one can generate collective 'goods' by generalising from one's own sense of respect for one's autonomy. Kant's view places a great deal of emphasis on the individual and that individual's autonomy. The assumption is that if everyone behaved in ways that respect the autonomous character of themselves and others, collective 'goods' would follow. Kant explored this assumption through the abstract concept of a perfect moral community, which he called the 'kingdom of ends'. Since, in the kingdom of ends, all citizens are involved in making all laws and those laws follow the categorical imperative (respect for the autonomy of the other), individual good, according to Kant, naturally becomes collective good. But just why should anyone accept this assumption? A conceptual framework for probing this issue arises from recent work on social contract theory.

3.2.3 Social contract theory

During the last 50 years there has been a resurgence of interest in social contract theory, and, hence, in the moral and political views of Hobbes, Locke, Rousseau and Kant. Probably the best-known contemporary social contract theorist is John Rawls. Amartya Sen (1992) has captured a widespread assessment of Rawls: 'By far the most influential – and I believe the most important – theory of justice to be presented in this century has been John Rawls's "justice as fairness"' (p. 75). The best known of his writings is *A Theory of Justice*. In this book, and in subsequent writings (1957, 1958, 1971, 1975, 1980, 1983), Rawls provides a solid, though controversial, attempt to establish the moral and political foundations of the social organisation of a liberal democracy. In essence, Rawls accepts the Hobbesian 'original position', of individuals – a kind of pre-social war of all against all, in which life was 'solitary, poor, nasty, brutish, and short' (see Hobbes, 1651). Against this background of individual rational self-interest (encompassing autonomy and respect for oneself), Rawls uses a game-theoretic structure. In order to ensure that individuals in the original position have no 'vested' interests, Rawls put them behind a 'veil of ignorance'. This veil hides from individuals all knowledge of their status, ethnicity, gender, etc. The central question for an individual, in this game-theoretic situation, is what general rules would a rational self-interested person promote for society? Since the person does not know what his or her gender, ethnicity, social status, etc., will be when the veil is lifted and the person is promoting rational self-interest, the rules will exhibit fairness and respect for oneself and others (as Kantian theory requires). To suggest, for example, from behind the veil of ignorance, that individuals of colour should be slaves for Europeans, or that women should not be permitted to vote or hold political office, is not a promotion, in a society, of one's own rational self-interest and does not promote respect for oneself, since when the veil is lifted one might be a female of colour. The result, according to Rawls, will be that individuals will not choose the principle of utility that underlies utilitarianism but will choose a concept of justice that is defined by two principles:

(1) Each person has an equal right to a fully adequate scheme of equal basic rights and liberties compatible with a similar scheme for everyone.
(2) Social and economic inequalities are to be arranged so that they are:
    (i) attached to positions and offices open to all under conditions of fair equality of opportunity
    (ii) to the greatest benefit of the least advantaged.

This formulation is from *A Theory of Justice*. In his Tanner lectures (1982), he gives a slightly revised version.

This version of social contract theory does not lead to a society in which everyone earns the same income, or has the same expectations or opportunities. It does require, however, that the least well-off person be as well off as the least well-off person could be in any alternative social arrangement (a maximin rule). It also requires that differences in income, expectation and opportunities not be a result of an unfair social arrangement but rather the result of the particular characteristics and choices of the individual in a society that is based on fair social rules. Fair social rules are those that respect individuals as autonomous agents who warrant equality of respect. It is this that the original position and veil of ignorance strategy guarantee to be the case. Once this ideal social structure is adopted, special rules to deal with the physically and mentally ill and non-compliant individuals can be formed.

Rawls' social contract theory, like other theories discussed in this section, is considerably more complex than can be described here and whether he has succeeded in providing a robust social contract theory is still the subject of much controversy (see Arrow, 1973; Daniels, 1975; Harsanyi, 1975; Sen, 1992).

3.2.4 Utilitarianism

Now I turn to the most influential consequentialist theory, utilitarianism. Although some aspects of utilitarianism can be found in writings before the publication of Jeremy Bentham's *The Principles of Morals and Legislation* (Bentham, 1789), it is Bentham (1748–1832) who is widely regarded as providing the first comprehensive theory based on utility. The best-known exponent is John Stuart Mill (1806–1873), and his formulation is the most widely cited (Mill, 1863). The name utilitarianism comes from the priority given to utility: utility must be maximised. The utilitarian maxim is that one should always act such that the greatest utility for the greatest number is achieved. The most common contemporary expression uses happiness as the relevant utility; hence, one should always act such that the greatest happiness for the greatest number is achieved. Other formulations use 'pleasure', 'good' or 'satisfaction of interests' as the relevant utility.

The superficial simplicity of this view is deceptive, as a closer look at 'happiness' will make clear. Happiness, in this theory, is not a vague, euphoric state of mind; it is a rich concept deeply connected to the Enlightenment idea of well-being (quality of life). As such, happiness (quality of life) can be measured through its essential components such as life expectancy, education, access to health care, wealth, leisure time and the like. This is a social concept. The quality of life of individuals depends intimately on the nature of the social arrangements. Hence, the injunction, 'one should always act such that the greatest happiness for the greatest number is achieved', requires that individuals act in ways that result in social structures in which the greatest happiness (quality of life) for the greatest number is achieved. Acting to increase access to education and health care are examples of always acting such that the greatest happiness for the greatest number is achieved. The actions are those of individuals but the measure of an action's ethical acceptability is collective (societal): greatest happiness for the greatest number. That entails that in principle sometimes the right action for an individual might diminish that person's happiness; as long as it leads to the greatest happiness *for the greatest number*, it is the ethically right thing to do.

For a utilitarian, the improvement in the measurable quality of life indicators (literacy, life expectancy, infant mortality and so on) within a society is evidence that people, individually and collectively (governments, corporations and other NGOs), are acting such that the greatest happiness for the greatest number is achieved. The quality of life of individuals will vary but as long as the greatest happiness for the greatest number is being achieved, the members of, and groups in, the society are acting ethically. This raises another subtlety of this theory. Although improvement in key indicators suggests that things are moving in the right direction, ethically speaking, how can the achievement of the greatest happiness for the greatest number be assessed?

This assessment will be comparative and contextual. It will involve comparing different possible social arrangements to determine which one achieves the greatest happiness for the greatest number given current circumstances. There will be considerable room for disagreement on which social arrangements will achieve the greatest happiness for the greatest number. These disagreements, however, are resolved through rational deliberation; reasons will have to be given, arguments presented and empirical support provided. Even if agreement on the ethically best social arrangements can be achieved, the phrase, 'given current circumstances' will open up another complex set of considerations. As circumstances change, the currently impossible may become possible or the currently possible become impossible. This guarantees that there will be no specific social arrangement that will be abstractly and universally the ethical winner. Contrary to what this might at first suggest, this contextual character of the theory makes it a much more robust theory, one that is dynamic. Circumstances constraining actions to improve the quality of life can be identified and possible ways to eliminate or ameliorate the constraint sought; currently intractable advances can be identified. Analysis of constraints will identify factors that can be improved with concerted effort. Hence, there are significant ways in which this theory can guide improvement.

A few generic examples of constraints, with no priority to the order, will sharpen the points made so far. First, the theory might entail a set of social arrangements different from those that are currently in place, but it might not be clear what actions will achieve it. Second, circumstances change and with them the social arrangements that achieve the greatest happiness for the greatest number. For example, the reliance on fossil fuels has a limited future; a number of elements involved in quality of life have been improved over the last century because of the availability of fossil fuels. Today, however, it is widely, though not universally, held that continued use of fossil fuels in the current ways and in the current amounts will diminish some elements involved in quality of life. To take this into account, social arrangements will need to change. Third, elements involved in quality of life are interconnected in complex ways. Sometimes improving one (say, access to health care) diminishes another (say, education), as in a financially constrained environment. Sometimes, improving one indicator at the expense of another will over time bring about improvements in others. For example, investment in education at the expense of health care may, in the short term, lower life expectancy but in the longer term may improve both. These are largely empirical matters and the best empirical knowledge is what should be relied on. Since empirical knowledge changes over time, so will the decisions about how best to achieve the greatest happiness for the greatest number. Fourth, sometimes the physical environment imposes constraints and attempts can be made to relax or remove them. This is usually achieved through science and technology. These constraints highlight the contextual and fluid nature of right action in utilitarian theory.

As one would expect, deontologists hold that utilitarianism is flawed. In their view, the one glaring flaw is that the principle of utility sometimes entails that an action is right when our moral intuitions declare it wrong. Consider two classic thought experiments designed to expose this flaw. In the first, Bill is hunting deer. Through a clearing, he sees with his field glasses someone place a suitcase in his base camp (about 2 km away) and then unwind some wire. It soon becomes clear that the suitcase is an explosive device and the wire leads to a detonating device. There are 10 people asleep in the camp. Bill immediately heads for the camp. Minutes after he sets out, he sees Ann heading into the camp; in less than a minute she will step on the detonating device and set off the explosives. If that happens, 10 people will die. Bill has just enough time, and is within range, to shoot Ann and avoid her stepping on the detonating device, but he is too far away to be heard by Ann. Should he shoot her? If 'one should always act such that the greatest happiness for the greatest number is achieved', then it appears he should shoot her: one life lost for 10 saved. But, from a deontological perspective, that is counterintuitive; Ann is innocent and has a right to live.

In the second case, Alex walks into the local hospital to visit his wife and newly born daughter. The physicians at the hospital are distraught because there are five people about to die; they need organ transplants and no organs are available. Each needs a different organ. They know that Alex, who is completely healthy, could be overpowered, sedated and anaesthetised. His organs could then be removed and the five transplants undertaken. If 'one should always act such that the greatest happiness for the greatest number is achieved', it appears that they should overpower him: one life lost, five lives saved. But intuitively that course of action seems grossly immoral.

The thought experiments are designed to show that there are moral principles that take precedence over consequences. That is, deontology trumps utilitarianism. Again, not surprisingly, utilitarians have responses to this challenge. At the core of one species of response is that these are bizarre cases falling at the edge of our experience. As such, no ethical theory provides a compelling basis for action. These cases take us beyond the limits of reliable intuitions or calculations; thankfully, we face them infrequently. Another response is to bite the bullet (punning aside) and reject the utility of moral intuitions as being reliable. Intuitions about matters of fact and ethical action are after all environmentally conditioned. A moment's reflection on intuitions about gay individuals or masturbation during the first half of the twentieth century compared with those today should make us wary of relying too heavily on our intuitions. Sacrificing one person (often to appease or curry favour with a deity or spirit) to benefit the whole has been deemed intuitively right in many societies in the past. Intuitions do not seem up to the task of sorting out moral dilemmas such as those just sketched.

A third species of response emphasises that the scope of the examples is exceedingly limited. There are numerous additional details and consequences that could and should be added to any situation. Were the physicians to overpower Alex, for example, this action would undermine confidence in hospitals and the health-care system, thereby undermining achievement of the greatest happiness for the greatest number. The explosives may be powerful enough that Ann will die anyway or the person who planted the explosives may still be lurking around and fully intend to detonate them himself if Ann does not, in which case, shooting Ann will not save the others. And on it goes. On the surface, this seems to suggest that a utilitarian will not be able to generate the right course of action unless all the information pertaining to the case is known. That is a tall order that will seldom be achieved and, possibly worse, it is not clear how one would determine that all the information was known. That suggests that this response rescues the theory at the expense of making it irrelevant to any actual determination of right action. Hence, as it stands, this line of defence is unhelpful. There is, however, a way of tightening up the essence of this response.

I have emphasised that utilitarianism, although based on individual right action, is in essence a theory about outcomes for a society: the greatest happiness for the greatest number. I have also indicated that happiness is understood in terms of quality of life and that these are, for the most part, measurable within a society. The problem with the thought experiments is not only that more information is essential; it is that the situation is divorced from the larger social context. Sacrificing every sixth person in order to have a sufficient quantity of organs for transplantation may increase life expectancy but will almost certainly lower individual liberty (the ability to realise one's interests). The task is one of balancing different elements of the social arrangements to maximise happiness for the greatest number of people within current constraints. Actions that improve social indicators of quality of life (happiness) in an aggregate assessment are right actions; this includes finding ways to relax or remove constraints that are impediments to such improvements. Decisions in bizarre circumstances are best based on the social arrangements (institutions, laws, policies, regulations, services and the like) that, as a whole, achieve the greatest happiness for the greatest number.

Utilitarianism understood in this way is robust and not open to the deontologist's critique. That utilitarianism ultimately requires a balancing of benefits and disutility explains why so many who take risk assessment seriously find utilitarianism an attractive theory. Even this brief survey of the theory makes clear that this is not, as it is often characterised, a theory in which people are treated as only means and not ends. The welfare of individuals lies at the heart of the theory. It recognises, as some other theories do not, that treating people as ends requires making trade-offs; and it recognises that maximising the interests of individuals ultimately depends on the social arrangements, and it is the entire set of social arrangements that must be judged. In this respect Rawls' theory intersects with utilitarianism, one indication that the frequently employed categories into which theories are slotted are inadequate and that there are many fuzzy boundaries between theories.

3.2.5 Ethical naturalism

As indicated above, I am persuaded by a number of factors that a naturalised social contract theory is the best ethical theory. An important factor for me is that social contract theory meshes more naturally with contemporary biological evolutionary theory. It assumes, for example, that rational self-interest is an essential component of ethical theorising just as it is an essential component of biological evolution. Utilitarianism, on the other hand, entails that, in principle, sometimes the right action for an individual might diminish that person's happiness (well-being); but as long as it leads to the greatest happiness *for the greatest number*, it is the ethically right thing to do. There are many ways to make utilitarianism compatible with biological evolution and with the pursuit of rational self-interest, but I find them less natural than social contract theory. Hence, I am not a utilitarian. Nonetheless, I accept that it is a powerful, robust ethical theory that cannot be dismissed easily, and it has had an enormous influence over the evolution of social arrangements during the last two centuries. I am considerably less optimistic that natural law can be made compatible with biological evolutionary theory and the pursuit of rational self-interest. Indeed, as will emerge in Section 4.2, I am not optimistic that it can provide ethical illumination in the twenty-first century.

I am an ethical naturalist and a social contractarian, but just what is naturalism, and, perhaps more importantly, how does it differ from natural law theory given they both appeal to nature? Like natural law theory, naturalism rests on the conviction that the proper basis for ethics is 'the nature of things', but that is where all similarity ends. For naturalists, ethics and natural science are inextricably connected; what is right and wrong, what one ought –

or ought not – to do is not based on self-evident principles, moral intuitions, maxims or principles but on the structure and requirements of nature (understood in the same way as physics, chemistry and biology understand nature). This stance seems, quite clearly, to ignore Hume's 'is-ought barrier' and to commit Moore's 'naturalistic fallacy'. It will turn out that it does not ignore Hume's distinction but it does commit the naturalistic fallacy. The latter is not surprising since Moore held that ethical attributes, such as good and right, are non-natural properties, and that is fundamentally what naturalism denies.

Let's begin with a fairly unadorned, but succinct, statement of a version of ethical naturalism found in Edward O. Wilson's *Sociobiology: The New Synthesis* (Wilson, 1973) and one that set off a storm of controversy:

> Camus said that the only serious philosophical question is suicide. That is wrong even in the strict sense intended. The biologist, who is concerned with questions of physiology and evolutionary history, realizes that self-knowledge is constrained and shaped by the emotional control centres in the hypothalamus and limbic system of the brain. These centers flood our consciousness with all the emotions – hate, love, guilt, fear, and others – that are consulted by ethical philosophers who wish to intuit the standards of good and evil. What, we are then compelled to ask, made the hypothalamus and limbic system? They evolved by natural selection. (p. 3)

For this to be a defensible account of ethics, it needs to be much more sophisticated. In essence, Wilson is claiming that the hypothalamus and the limbic system are the result of evolution by natural selection. They are also the locus of human emotions – disgust, hate, love, etc. Human emotions are, therefore, no more than the response of an organism to its environment, a response that is a function of its natural biological origins. Moral intuitions about good and evil, right and wrong, are merely a reflection of these emotions. The immediate response to Moore arising from this thesis is that the open question is illegitimate. Moore's claim was that although a statement, 'individuals despise those that despise them and love and help those that love and help them', may be empirically established, one, nonetheless, can ask of the behaviour, 'is it good?' Wilson's claim is that natural selection has left us no choice, so the question is irrelevant. If a defender of Moore responds that the question is legitimate and being human requires we pose and answer it, Wilson would no doubt point out that we address it by reference to the very evolved emotions (intuitions) that gave rise to the behaviour. In this case, what progress have we made over the initial statement? None; a set of responses to environmental circumstances evolved and so did our intuitions that they are the right responses. It makes sense within evolutionary theory for those responses to have evolved (because they are survival enhancing) and also for intuitions to have evolved, at the same time, that make us believe the responses are morally right.

This reasoning highlights the importance of biological evolution to ethics but there are some missing elements. For Wilson's account to succeed, there needs to be a fairly tight connection between the environmental circumstances and the response; environmental circumstance X triggers response Y and that is the best response in terms of enhancing the differential reproductive success of the individuals exhibiting it. Also, responses have to be unmediated or at least only weakly mediated. That is, the responses have to be reasonably automatic. Neither of these requirements, however, is met in humans. Many physiological responses are more or less automatic and are usually maximally survival enhancing (e.g. an immune response to a virus), but most behavioural responses are not. Environmental circumstances are usually complex and a range of responses is possible. Cognition allows us to deliberate on the most appropriate response; so, responses are not automatic. We may have a propensity to behave in a certain way in response to a class of circumstances but we usually can override that propensity. Indeed, the survival success of humans to this point has been due to our cognitive ability to decide to suppress our propensities and to respond differently based on the assessment of circumstances. We can recognise subtle differences between circumstances and shape our responses accordingly. These factors suggest that humans can, and do, evaluate circumstances and responses (courses of action). That deliberative process is what ethical theories purport to address. It is, of course, constrained by our evolutionary past – emotions loom large in the deliberative process – but it is not determined by that past. We have even found ways to diminish the impact of a raw emotional response to a circumstance; we have been able to understand evolutionary dynamics and judge that a response – even a physiological response – is, in the current circumstance, not the best response even from an evolutionary perspective. So, contrary to what Wilson's reasoning suggests, there is a lot of room for deliberation on appropriate courses of action. The important task is discovering the underpinnings of that deliberative process, and specifically, in a social context – that is, a context in which a group of people are coexisting.

So, clearly, Wilson's version of naturalistic ethics in *Sociobiology* is flawed but that does not mean that naturalism is untenable. A central premise of Wilson's naturalism is that the touchstone for ethical theorising and reasoning is evolutionary theory – evolutionary processes and patterns. Any ethical theory, injunction or principle that is inconsistent with what we know about how evolution has shaped us and how evolutionary processes operate today must be rejected. This premise and the constraint it places on ethical theorising seem correct. It is analogous to claiming that any recommended behaviour that is inconsistent with what is known from physiology must be rejected. For example, a claim that it is good for everyone to have a litre of blood removed every three days is inconsistent with current physiological knowledge and, hence, must be rejected on that ground. This, however, is a negative thesis; it poses a test that an ethical theory, and the injunctions and principles it generates, must pass. As such, it does not, on the surface, seem to be a promising candidate for **constructing** an ethical theory and providing a justification for accepting it. In short, a naturalistic ethics, which entails ethical injunctions and principles – that is, an ethical theory generated and justified by reference to natural science alone – does not seem possible. This, however, underestimates the robustness of the naturalistic approach.

A scientific theory entails statements about the behaviour of things under this or that set of circumstances. Sometimes the behaviour has already occurred and the entailed statement explains why it occurred; at other times the entailed statement predicts what will happen if this or that set of circumstances occurs. Similarly, sometimes an ethical theory entails statements about how one should behave in this or that set of circumstances; at other times it entails statements that are used to assess a behaviour that has occurred. These statements are norms and these norms regulate human social behaviour. For a number of the theories we have canvassed in this section, the goal has been the entailment of norms that produce a functioning society. The UN Declaration is aiming to achieve 'freedom, justice and peace in the world'. These concepts need to be fleshed out a bit, but in Western countries there is a shared understanding of them. Achieving this goal is dependent upon 'the recognition of the inherent dignity and of the equal and inalienable rights [such as the right to life, liberty and security of person] of all members of the human family'. An ethical theory can now be constructed to embody the goals and norms. In this case, social contract theory arises naturally from the goals and norms.

A naturalistic ethical theory based on biological evolution will begin by examining the evolution of social arrangements. By that examination, norms will be identified, norms that have increased the survival of the individuals in the society as well as ensuring the survival of the social framework that makes individual survival possible (the social framework must be sustainable over time). Some social arrangements will enhance individual survival more than others; those social arrangements will also survive. For a naturalist, acceptance of a social framework known not to enhance the survival of the individuals within it or to be inferior in achieving that goal is irrational. So the fundamental question is, 'Why is individual survival the goal?' The answer is, because, if anything qualifies as a self-evident truth, the goal of individual survival seems to be it. Denying it seems entirely irrational. To an evolutionary biologist, that is just the way things are in the biological world. It is the reason we have the characteristics we do, including our desires. Furthermore, not striving for this goal will result in extinction of the species; any ethical theory that entails that individual demise and human species extinction is a good thing would be a bizarre theory indeed and evolutionary biology explains why.

What kind of social arrangements will maximise the achievement of this goal? Evolutionary theory takes us a long way towards the answer (Thompson, 2002). In essence, one is looking for social arrangements that create a stable society. Social collapse, which will be the fate of chaotic, dysfunctional societies, will not maximise the survival of the individuals in that society – quite the opposite. The evidence from evolutionary biology – and other fields such as anthropology and sociology – is that social arrangements that foster social cooperation are essential. Now, we are up and running in the development of a naturalistic ethics; we may not always get the arrangements right but that is what we are striving to achieve, and we adjust and correct over time, as we have done for more than 50,000 years. Evolutionary biology and other fields also suggest that fostering social cooperation will involve maximising the satisfaction of individual interests; in turn, this will require norms such as rights (probably inalienable rights such as right to life, liberty and security of person), justice, equality before the law and so on. This takes us close to social contract theory (and close to the UN Declaration). All this is achieved by appealing to standard canons of reasoning and only empirical science. That is what makes this a thoroughly naturalistic theory; all the ethical norms and resulting theory follow from the examination of nature, and nature alone.

3.3 Harm and risk analysis

The concept of harm is exceedingly difficult to define with precision. Broadly, it is defined in terms of negatively interfering with the achievement or satisfaction of the interests of another person. This is too broad in several respects. First, 'interests', *simpliciter*, in this definition, is too broad. A more nuanced definition in ethical, social and political philosophy circumscribes interests to legitimate interests. If, for example, Nadine interferes with Bill's interest in robbing a bank, that is not a harm even though it thwarts one of Bill's interests. Hence, Joel Feinberg (1973) defines harm as, 'the thwarting, invading, defeating or setting back of *legitimate* interests' (emphasis added).

Second, some harms are idiosyncratic, even frivolous. Describing a specific harm as idiosyncratic or frivolous is, of course, contestable. Many harms that are reasonably described as idiosyncratic or frivolous are based on subjective emotional responses such as making someone feel uncomfortable, insulted, offended or annoyed. If Camilla chooses to wear a thong on a beach, it might make others on the beach uncomfortable; they might even find it offensive. Does that make it a harm? Has Camilla harmed someone? Judgements will differ. She likely has invaded their interest in having a relaxing day at the beach. Some might consider her action inconsiderate; others will consider those taking offence to be intolerant – forcing their purely social (prudish) mores onto others. After all, in the early twentieth century even men were mostly covered up on the beach – times change. In cases like these, people's subjective responses will vary – often widely. The challenge in sorting this out is that if legitimate harm extends to cases of offence, the harms one can inflict (frequently without intent or even knowledge) expand dramatically. More problematically, the permissible actions a person could perform would be excessively constrained.

Third, harms cannot be determined in isolation. If Gordon, who is a nudist, wants to construct a privacy fence on his property and it results in the frustration of an interest of his neighbour Samuel because it blocks his view of a favourite tree, has a legitimate harm occurred? Satisfying Samuel's interest in being able to see the tree will result in thwarting Gordon's interest in privacy. So someone is going to be harmed whatever happens. Most cases will involve competing interests of this kind; inevitably someone's interest will be frustrated. Some harms may be proscribed entirely – killing, maiming, torturing and enslaving, for example. Hence, if satisfying an otherwise legitimate interest requires inflicting one of those kinds of harms on someone, it must remain a frustrated interest. These proscribed harms are often embedded in documents guaranteeing fundamental rights and freedoms. For the rest, the entire social fabric – especially laws – constitutes an attempt to **balance** legitimate interests and harms. That is the social challenge: to balance competing interests and harms in ways that are deemed just and fair.

In law, 'interests' is usually modified by 'legal' or 'legally protected', as in the US *Restatement of the Law, Second, Torts*: 'Harm is the invasion of legally protected interests' (American Law Institute, 1965–79). The term 'tort' is derived, via French, from the Latin word *tortus* (literally, 'twisted'), and it means a breach of duty. Also, in law the phrase 'interfering with the achievement or satisfaction of interests' is captured by the term 'injury'.

As we have already seen, Immanuel Kant's famous dictum (*Groundwork of the Metaphysics of Morals*, 1785), 'Act always so that you treat humanity, in your own person or another, never merely as a means but also at the same time as an end in itself', is an echo of an older phrasing – one found in the Bible (Luke 6:31), but found also in even earlier writings – 'And as ye would that men should do to you, do ye also to them likewise' (King James Version). This is frequently rendered, 'Do unto others as you would have them do unto you.' Being treated as a person (an end) involves respecting an individual's legitimate pursuit of her interests. Not doing so gives rise to a harm/injury. Such harm/injury (whether intentional or through negligence) requires redress – frequently through compensation. The redress is an attempt to undo the harm and/or punish the perpetrator. For example, someone awarded compensation for damage to personal property can use the compensation to repair the damage *and* the perpetrator will be punished monetarily by the loss of funds – a kind of transfer of injury from the victim to the perpetrator. In some cases, the actual injury cannot be rectified (such as the loss of a limb), but a surrogate rectification can usually be found, even if it is money rather than a limb.

The discussion of harms to this point has focused mostly on individuals being harmed by having the pursuit of their legitimate interests frustrated by other individuals or a small group of individuals. The discussion now has to be expanded to include actions of governments, institutions and industries, because these are the most germane to agricultural biotechnology. Also, the scope and constraints on 'legitimate' needs closer attention since not all interests are legitimate. Furthermore, as we have noted, the pursuit of one person's interests will often compete with the interests of others; agricultural biotechnology is debated in contexts where these competing interests abound.

When a government or one of its agencies approves, declines to approve, or withdraws approval of a product for use as a medicine or food or in an industrial process, it confers a benefit on some individuals (collectively or separately). Sometimes the benefit is to the consumer, as in the approval of a new, effective pharmaceutical, or in the rejection or recalling of a product believed to be a toxin, pathogen or carcinogen or to be contaminated. The rejection or withdrawal of a product, however, is very likely to result in harms to other individuals such as employees of an industry who may be laid off or shareholders whose investments decline in value. The justification for proceeding nonetheless with the rejection or withdrawal will be that harm to these individuals is outweighed by the benefit (in this case, avoidance of harm) to others.

Sometimes the benefit is to an industry and the consumer, as is the case with approvals of pharmaceuticals. There may be known harms (often in the form of potential side effects), but the approval is justified because the benefit to the consumer is deemed to outweigh the harm. The benefit to the pharmaceutical company is profit. In some cases, not approving a pharmaceutical will harm one class of persons and approval will harm a different class of people. The justification for one course of action over another involves assessing the balance of harm to benefit. Sometimes the benefit and the harm are to a single individual, as in the case of relief of arthritic pain with a pharmaceutical (e.g. Vioxx) whose long-term use has negative effects on the cardiovascular system, the kidneys, the liver and so on. Sometimes one group of people realise the benefit and a different group the harm, as in the case of gravel extraction for construction, which benefits those for whom the construction is undertaken but harms those who live near the truck route at the gravel pit. Sometimes all members of a society realise the benefit (e.g. mobility by automobiles, trains, planes) and the harm (climate change). Very few benefits come without actual harms or the potential for attendant harms. The determination of benefits and harms and assessing the relative balance is known as risk assessment; risk assessment involves determining and balancing *the probability and magnitude of benefits* and *the probability and magnitude of unwanted harms* (imminent or future).

Hence, four variables dominate the assessment of risk and the reasonableness of taking that risk:

- magnitude of the harm (M(h))
- probability of the harm occurring (Pr(h))
- magnitude of the benefit (M(b))
- probability of realising the benefit (Pr(b)).

As a general rule, taking a risk is only reasonable if:

$$M(b) \times Pr(b) > M(h) \times Pr(h)$$

This is a minimum threshold; if $M(h) \times Pr(h) > M(b) \times Pr(b)$, it is almost always unreasonable to take the risk, but the converse is not the case. Sometimes the magnitude of the potential harm is horrific even though the probability of its occurring is very low. To take such a chance for a highly probable but modest benefit might well be unreasonable. In addition to these four variables, there is at least one other factor that must be considered: is there another course of action that will achieve the benefit with a lower magnitude or probability of harm?

Joel Feinberg (1973) has expressed these elements of risk analysis in slightly different terms:

- value of a desired outcome
- probability of the desired outcome
- probability of harm in securing the outcome
- severity of the harm
- alternative methods of achieving the outcome with a lower probability of harm or less severe harm.

This is a richer formulation because it introduces an evaluative element. Obviously, in the first formulation, the magnitude of a benefit is based on its value as a benefit and the location on a magnitude scale; by being explicit about the connection between value and magnitude, as Feinberg is, there is less chance that the magnitude of benefit will be mistakenly treated as an 'objective' measure. After all, magnitudes are quantities and quantification suggests, falsely in this case, objective measures and standards.

Something needs to be added to both schemata. Suppose an action has a high probability of bringing about a large amount of benefit with a very low probability of harm and, should harm occur, it will be minor. It still might not be reasonable to engage in the action; the action, for example, might foreclose achieving a benefit with a higher value. This is frequently the case with an economic 'zero-sum game' – a game in which any gain in one part of the system results in an unavoidable and equal loss in another part of the system. For example, the gun registry system in Canada (implemented in 1998) was expensive to implement and is expensive to maintain. The cost remains elusive as this quotation from the Canadian Broadcasting Corporation (CBC) illustrates:

> Canada's controversial gun registry is costing taxpayers far more than previously reported, CBC News has learned. Nearly $2 billion has either been spent on or committed to the federal program since it was introduced in the mid-1990s, according to documents obtained by Zone Libre of CBC's French news service. The figure is roughly twice as much as an official government estimate that caused an uproar across the country. The gun registry was originally supposed to cost less than $2 million. In December 2002, Auditor General Sheila Fraser revealed that the program would run up bills of at least $1 billion by 2005. But the calculations remained incomplete, so CBC News obtained documents through the Access to Information Act and crunched the numbers.

The most reasonable and defensible numbers at this point are CAN$1.2 billion in implementation and CAN$85 million in annual operating costs.

I support gun control but I, along with many others, am sceptical that the registry has added any benefit over existing gun-control measures, which I support. Those pre-1998 measures severely restricted ownership of handguns, both automatic and semi-automatic weapons. Ownership of long guns is controlled, not restricted, and owners must have a photo-identification permit to possess a long gun and to purchase ammunition. Obtaining this licence requires a police background check (including contacting neighbours). To buy a long gun requires a different permit, which in addition to a background check requires proof of firearms training. These have been effective measures and most Canadians support them. The long gun registry was in addition to these existing measures. Notwithstanding the scepticism about its benefit, let's assume there has been a benefit of 40 lives saved a year;[2] let's also assume

[2] Statistics Canada recently reported, 'There were 200 homicides committed with a firearm in 2008, 12 more than in 2007. The rate of homicides committed with a firearm has increased 24% since 2002.' 2002 is four years after the gun registry was implemented. No one expected that the registry would end homicides committed with firearms, but, given

that the only direct harm is minor inconvenience (forms to be submitted, small fees to be paid and the like). Not surprisingly, many gun owners and their associations hotly contest this minimising of harm. That is what makes this a robust case; it is real, and manifests all the complexity of benefit-harm assessment where differing values are in play and the interpretation of 'data' varies widely. It mirrors the complexity of the agricultural biotechnology debate, in which so many voices on either side want to reduce the issues to simplistic platitudes and to information highly crafted to seem correct and obvious.

Focusing again on the zero-sum game character of this, if the CAN$1.2 billion were used to institute suicide prevention programmes and the CAN$85 million annual cost was used to sustain those programmes, many lives could have been saved. Statistics Canada (Statistics Canada, 2003, 2006) reported the number of suicides in 2000 to be 3,600 people (2,798 males, 807 females) and in 2003 to be 3,764 (2,902 male, 862 female). So, the number hovers around 3,700 (a suicide rate of 11 per 100,000). No programme of prevention is going to be completely effective, so suppose 1 in 50 are prevented each year – a low success rate, for sure. In that case, about 74 lives per year would be saved. This is where values enter; if death reduction is the focus, suicide prevention wins hands down. If a specific kind of death reduction is the focus, then the winner will depend on which kind is valued. If the death rate is not really the true focus but rather something like public confidence in individual safety, then the gun registry will likely trump suicide prevention. Even this, however, is not straightforward; there may be other much more effective ways to achieve public confidence in individual safety if CAN$1.2 billion start-up and CAN$85 million annually is available, such as increased police presence by hiring more officers. The evidence suggests that this is, in fact, a better investment to improve public safety.

This example illustrates how interconnected are the outcomes of actions and how remarkably challenging is the process of reasoning to the best courses of action. As one might expect with all these competing interests, advocacy groups abound. Advocacy groups, by their nature, attempt to promote their common interests over those of others. This marketplace of public discourse

the cost, it is reasonable to assume that those championing it expected some decline, not an increase of 24 per cent, which is a rate relative to population. A 20 per cent decrease seems reasonable, so approximately 40 'saved' lives is a reasonable number to use in this example.

and advocacy is a hallmark of a healthy democracy but it also increases the complexity of social decision-making. Two features, however, of the tactics of advocacy groups that do not advance democratic decision-making are the over-simplification of issues and emotive language and images. Successful advocacy groups know well that the simpler, more sharply focused, and more concise the message and information are, the higher the probability of garnering public support. Regrettably, most important issues are complex; the evidence is sometimes not decisive, the perspectives less coherent than portrayed and the potential resolutions illusive and/or obscure. Part of the simplification involves removing these grey areas. A message is cleaner and clearer if it can be structured in either-or terms; you are for us or against us, you are part of the solution or part of the problem. There is no middle ground and no grey area; things are black or white. Advocacy groups thrive on framing their message as, 'our way or the wrong way'. It is rare, however, that issues bifurcate neatly in this way; there are usually lots of ambiguity, multiple alternatives and considerable uncertainty about evidence, appropriateness and probability of success of proffered solutions, ulterior motives, and so on. In addition to the tactic of simplification, the more emotional attachment an advocacy group can elicit for its cause, the greater its success in attracting converts (donors, volunteers and political allies). Pictures of seals being clubbed to death are emotionally powerful. Unfortunately, they mask the complexity of the issues involved, and it is the issues, not the emotional (gut) reaction, that public debate and decision-making should be about.

Regrettably, the methods of Western democratic governments have become, of late, more like those of advocacy groups; they offer messages that are simple, concise and bereft of information. And, on the issue of information, far too many democratic governments have resorted to hypersecrecy – under the guise of national security (which is sometimes but rarely a genuine justification) or some other smokescreen – thereby stifling informed, deep and meaningful debate and public discourse, all to the serious detriment of a true democracy. When people are well informed and have the opportunity to consider the positions and arguments of others, the outcome of the electoral process is very likely to represent the considered will of the populace. When information is tightly controlled and legitimate voices are muted, an electoral process descends to the level of a popularity contest.

This kind of strategy by advocacy groups and governments often creates public awareness, but it does not result in healthy public discourse and definitely does not provide a rational basis for developing social policy. For that, rational methodologies for weighing competing interests and balancing them are needed. Ideally, those methodologies will result in a society (public structures and policies) that maximises the satisfaction of individual interests and minimises the thwarting of those interests. There will be different methodologies to address different contexts of decision-making, although they, of course, must collectively be consistent. The methods of legal reasoning and decision-making are clearly an appropriate way to protect constitutionally entrenched rights and freedoms, and are useful in settling an array of other cases of conflicting rights.

In the context of agricultural biotechnology, deliberations and decisions for the most part (patents being one exception) centre on avoiding, mitigating and managing potential causes of personal harm – specifically, harm to individual health and well-being – and environmental harms. The appropriate methodology for this is risk determination, assessment and maximum mitigation. To repeat what I take as obvious, the potential for harm is ubiquitous and not eliminable. To pretend otherwise is Pollyannaish. Hence, the goal of a rational methodology, in the context of the potential of harm, is identification, assessment, maximum mitigation and continual, effective monitoring.

The Institute for Risk Research (IRR; University of Waterloo, Canada) (Nathwani *et al.*, 1997; see also Lind *et al.*, 1993) has developed an approach to developing a rational, quantitative decision-making methodology. Although many other methodologies have been advanced, the core of most is similar, and the IRR methodology is an exemplar. Its approach begins by articulating the nature and scope of the basic individual interests on which all others depend. It encapsulates these in three common assumptions:

1. Long life in good health with few restrictions on individual choice is a fundamental value.
2. Risk mitigation that does not increase item 1 is deemed to detract from it and cannot be justified.
3. Benefits and harms must be reasonably distributed.

It also assumes that any rational risk assessment and risk management methodology will adhere to the following four principles:

1. The Accountability Principle
   Decisions for the public in regard to health and safety must be open, quantified, defensible, consistent and apply across the complete range of hazards to life.
2. The Principle of Maximum Net Benefit
   Risks shall be managed to maximise the total expected net benefit to society.
3. The Kaldor–Hicks Compensation Principle
   A policy is to be judged socially beneficial if the gainers receive enough benefits that they can compensate the losers fully and still have some net gain left over (i.e. losers can be transformed into non-losers with some residual gain for the gainers).
4. The Life Measure Principle
   The measure of health and safety benefit is the expectancy of life in good health (Nathwani *et al.*, 1997).

In light of the initial three assumptions, one would have expected a principle related to maximising liberty – more on this later. Principles 1 and 2 seem straightforward – 'net benefit to society' entails that the benefit to each individual is maximised consistent with a functioning society. Principle 3 needs some unpacking. Consider the situation where a large multinational industry wishes to locate a processing plant on a river at a point where it is wide and slow moving. An ideal location is found but a small village is located there already. The population of the village is 300. The processing plant and the village cannot coexist because daily the plant releases a non-toxic (already determined by the relevant government environmental agency) but foul-smelling vapour. The financial return that the industry will realise from this plant – especially in this ideal location – is around US$3 billion annually. In accordance with the Kalder–Hicks compensation principle, the industry makes an offer to the citizens. The industry will build a new village 200 km upstream (new houses, new roads, a recreation centre and so on). In addition, each citizen will receive a cash payment of US$1 million net of taxation (a gross payout of approximately US$1.4 million per citizen). The new village will cost US$600 million to construct and the cash payout to the citizens will be US$420 million. This compensation will reduce the net profit for one year by US$1.2 billion. The initial losers – the citizens – have now become non-losers (perhaps even beneficiaries).

The Kalder–Hicks compensation principle is embedded in Pareto optimality models (named after Vilfredo Federico Damaso Pareto). A set of social (or strictly economic) arrangements can be Pareto inefficient (suboptimal) or Pareto efficient (optimal). An arrangement is Pareto inefficient if an individual or group of individuals can achieve a benefit without any loss on the part of other individuals. That outcome indicates that the original arrangement did not maximise benefits since additional benefits could be extracted from the system without anyone losing. Such a system can, and should, undergo Pareto improvement to wring maximum benefits out of the system. A set of arrangements is Pareto efficient, or optimal, if no individual can achieve an increased benefit without some other individual or individuals experiencing an equivalent decreased benefit; it, in effect, is a zero-sum game.

Of the large variety of ways in which a Pareto-inefficient system can be improved, the Kalder–Hicks compensation principle specifies one species of improvement, one in which increased benefits are realised but in a way constrained by a compensation system. An important feature to highlight, in the context of the IRR method of risk analysis, is that there has to be residual gain after the compensation. If fully compensating losers consumes all of the benefits realised by the initial gainers, there are no losers but also there are no winners. The winners used the winnings to compensate the losers. In such cases, there is no Pareto improvement; no additional benefit has been wrung from the system. This is important in risk analysis because if a benefit realised by some individuals equals the harm to some other individuals, there is no system-local reason to inflict the harm; system-local means focusing just on the relationships between the benefits and harms. There may be non-system-local reasons, such as a justice-based redistribution of social resources (social goods), some examples of which are given in the next few paragraphs.

An important caveat is included in the principle: the requirement that losers can be compensated does not entail that they have to be compensated. The Kalder–Hicks principle only requires compensation *in principle*. A Kalder–Hicks improvement can result in losers as long as the gainers realise enough additional benefit that they *could* fully compensate the losers. The Kalder–Hicks compensation principle identifies only those changes to a system that increase benefits. Compensation in principle permits changes that create losers who may or may not be compensated; the only requirement is that they could be compensated. One can imagine cases where the individual who benefits was disadvantaged by the original social arrangements and the loser had an abundance of benefit. On many concepts of social justice, no compensation is warranted. In such a case, there would be a Kalder–Hicks improvement even though compensation never occurred. There are also cases where the loss to the loser is minimal and, hence, falls below a threshold of concern – minor harm such as being required to get a booster vaccine at a local vaccination-only clinic instead of from a local physician who is physically closer. The benefit may be a reduction in health-care spending. The compensation might be putting a price on time and expenses for the loser to travel further. If this can be done with residual savings to the health-care system still being realised, this change will result in a Kalder–Hicks improvement even if the losers are never actually compensated. Hence, the compensation in principle ensures that the system is Pareto improved but allows for a realignment (redistribution) of social resources based on justice and fairness.

If every change that results in a benefit to someone also results in an equivalent loss to someone else, the system is Pareto optimal; no additional benefit within the system can be realised. Different distributions of the benefits are possible, and on other grounds may be desirable, but no additional benefits can be realised and so there is no possibility of Pareto improvement in that system; whatever reasons there may be for a redistribution of benefits, no Pareto improvement occurs because the system was in a Pareto-optimal state. Consider a social housing project. The government has determined that very low-income families require government-provided shelter. The evidence suggests that the most cost-effective way to do this is to build government-owned apartment blocks. Of course, the government's assumption and adduced evidence are always open to challenge. The government owns some land in a suburb of the city and moves forward with the project. The value of the existing real estate in the area decreases as a result of the project since potential buyers worry about a decline in the quality of the local schools, the peer groups with whom their children associate and so on. Hence, there are real gainers (those being given shelter) and there are losers. If the harm (losses) equals the benefit (gains), no Pareto improvement has occurred. This example is characteristic of a significant proportion of social decision-making and assessments of benefit and harm.

Consequently, that a set of social arrangements is Pareto optimal (efficient) or satisfies the Kalder–Hicks compensation principle does not entail that it is socially acceptable, just or even socially stable. Consider another example – an economic example. Imagine a set of social arrangements in which

85 per cent of the wealth is held by 10 per cent of the population. In addition, the system is Pareto optimal because any gain in the wealth of a member of the 90 per cent of impoverished people will result in a reduction of the wealth of one or more of the wealthy 10 per cent (as would a redistribution of the wealth among the wealthy 10 per cent). Also, someone in the impoverished 90 per cent could increase her wealth at the expense of someone else in the impoverished 90 per cent, but that increase would be marginal since each individual in the impoverished 90 per cent has only a small portion of the total wealth. The system is Pareto optimal but not a just social system on any accepted concept of justice. It is also likely that it is not socially sustainable; it is a recipe for revolution. In addition, it is highly likely that it is sufficiently unacceptable to the majority of the society that only the use of brute force could maintain such a system. This particular Pareto-optimal social structure is one member of a set of Pareto-optimal social structures. At least one member of that set will satisfy Rawls' concept of a just society (and be a socially stable and acceptable one), namely, the set of social arrangements that makes the worst-off person better off than the worst-off person is in any of the Pareto-optimal alternatives.

One fundamental element of a social system that satisfies Rawls' concept of a just society is explicitly mentioned in the three common assumptions set out above, namely, liberty – the freedom to pursue one's interests. It is fundamental because it is the *sine qua non* (that without which nothing) of a social system that maximises the achievement of individual interests. As articulated earlier in this section, liberty, however, has limits; in a society, my pursuit of my interests needs to be reconciled with your pursuit of your interests, indeed, each individual's pursuit of his or her interests must be reconciled with every other individual's pursuit of her or his interests. That is why the assumption expresses liberty as, 'few restrictions on individual choice', making it clear that there will be restrictions but they must be minimised. Every restriction must be justified.

Having now explicated the assumptions and principles of this method of risk assessment and management, let's look at the central concept, that which the enterprise is attempting to maximise, namely, benefit. There, of course, is a plethora of potential benefits and whether something is a benefit is idiosyncratic. That is why individual liberty is so important; with defensible limitations, it is up to me to decide whether something is a benefit and which benefits I wish to pursue. There are, however, some benefits that everyone agrees are benefits and wishes society to promote; a long life and good health are two key ones[3] and are fundamental in the context of agricultural biotechnology. The vast majority of the debates are about the benefits of an adequate, safe supply of food, which is essential to health and longevity, and about the potential negative health and environmental effects. Sometimes the claimed harm is direct such as allergic reactions, sometime indirect such as environmental degradation, but virtually all get distilled down to threats to health and longevity. Wealth is another benefit since, for the most part, as wealth increases, an individual's potential to achieve more interests increases. Wealth is relevant to many issues in agricultural biotechnology but is less prominent than concerns about benefits and harms to health and longevity.

Risk assessment is both qualitative and quantitative. The value one places on a benefit or on avoidance of some harm is irreducibly qualitative, and that is why a defensible quantitative assessment must, at its core, rest on widely shared values. As indicated, good health, longevity and wealth do seem to be widely shared values – indeed nearly universal, since very few people will pursue ill health, short life or poverty. Fortunately, these widely pursued benefits are also quantifiable. Wealth is obviously quantifiable, although there may be some debate about the relevant components. A robust and widely used set of components is one's liquid assets (e.g. money in bank accounts, guaranteed income certificates), the value of real property (e.g. house), annual income (e.g. salary, investment income), the value of other property (e.g. furniture, art work) and debt. The sum of these (where debt is expressed as a negative asset) is a wealth index usually stated in a local currency or in US dollars. Credit card companies, institutions underwriting mortgages and the like employ this measure. So measures of wealth are readily available and are the basis for a significant amount of economic decision-making.

Expected longevity of individuals is also quantifiable and is the basis for decisions made by insurance companies. Life insurance companies have been calculating life expectancy for classes of individuals (e.g. smokers, those who engage in extreme sports, those with a personal or family history of health

[3] There may well be some people who wish to pursue ill health or a short life, but this will be a small minority. Even those who engage in activities that will potentially lead to ill health are not pursuing ill health; it is a side effect of something else they are pursuing. Those who attempt (or succeed in) suicide are arguably pursuing a short life, but, as already noted, 4,700 suicides per year in a population of approximately 34 million is a small minority. Even if one adds genuine but failed attempted suicides (in many cases, individuals who attempt suicide do not want it to succeed), we have something like 10,000; that is about 0.029 per cent.

problems, and those in midlife in good health) for a long time. It is the basis for setting premiums and hence has to be highly dependable. The mean (common average) of the individual life expectancies provides a social indicator of average life expectancy. Changes in the mean for the society or some subsets of it are used as indicators of progress or regression.

Good health is a bit more challenging. Maximum absence of illness and disease is what people seek but quantifying that is difficult. Probably the best measure is the number of days lost to ill health ('days lost' meaning days in which engaging in work and the entire range of leisure activities is not possible). This captures the idiosyncratic nature of ill health; one person's irritation may be another's debilitating pain. It also captures in a natural way the severity of ill health; the greater the number of days judged by the person to require missing work or some leisure activity, the more severe is the ill health. This, of course, rides roughshod over all kinds of subtle features of ill health but it incorporates many of the key ones. In addition, two things that are much easier to quantify correlate highly with good health understood in this way: wealth and education. Consequently, they can serve as surrogates for good health in many, but not all, contexts.

Drawing these threads together, we can fashion a quality of life index based on wealth and life expectancy. Let:

L = LQI (life quality index)
g = wealth
w = time spent producing it
e = life expectancy
$(1 - w)$ = time left not spent producing wealth (leisure time)

Then, the quality of life index is

$$L = g^{w} e^{(1-w)}$$

L is the product of wealth (weighted for time spent producing it) and life expectancy (weighted for leisure time).

This index can apply to an individual or a society. In the latter case, it is a social indicator (an indicator of the overall quality of life in that society); as the indicator improves, it indicates that the quality of life in that society has improved (on average) – benefit has increased – which should be the goal of all societies. If the indicator decreases, it indicates that the quality of life in that society has worsened – harm is being increased. If an action, product, process and the like results in a positive change in the product of g and *e*, the society and, hence, as an aggregate, the individuals in it have experienced increased benefit. Harm is the inverse of benefit. If some action, product, process, etc., results in a negative change in the product of g and *e*, harm (loss of benefit) has increased.

A change in *L* can be assessed as:

$$dL/L = w(dg/g) + (1 - w)de/e)$$

If g^w is negative (reduced, say, by the cost of cleaning up a major contaminant spill), then dL/L will be negative and harm has increased. Hence, for there to be a *net* benefit, dL/L must be positive. For dL/L to be positive, $dL/L > 0$; hence:

$$w(dg/g) + (1 - w)de/e > 0$$

Since $w + (1 - w) = 1$, only one measure is required; let that be *K*. Then

$$dg/g + Kde/e > 0$$

So, a net benefit occurs when $dg/g + Kde/e > 0$

L in this formulation does not address the question of distribution. That is, the life quality index may have increased because a few very rich people became even richer – maybe even at the expense of very poor people. To address this requires the incorporation of a distribution of benefits requirement; a principle of distributive justice needs to be included (see Kagen, 1998, pp. 48–54, for an excellent discussion).

3.4 The precautionary principle

As already stated, risks abound; few of life's activities do not involve risks. In light of this, the rational response, in the view of many people, is to identify risks, to assess the reasonableness of an action given its risks, and to manage risks associated with reasonable actions. Some people, however, have taken a different position. They argue that in many cases precaution is the rational course of action. On the surface, it is difficult to disagree that precaution is rational, and it seems indistinguishable from the mantra of identify, assess and manage risks. But some who espouse it under the label of the precautionary principle imbue it with special meaning.

The strongest version of the principle asserts: take no action unless it is certain that no harm will occur. This version is hyper-precaution and virtually no actions will be justifiable. Only few people and groups espouse a literal reading of the strong version. Most advocates of this strong version do not interpret it literally. It rather serves as an ideal – a signal that the hurdle of acceptable action is very high. It is, however, frequently for these individuals a rejection of the method of offsetting harms with benefits. The goal is to do no harm and individual or social actions must come as close as reasonably possible to that goal.

By contrast with this strong version, the moderate version advocates that actions should result in no unmitigated harm. One does not have to be certain that no harm will occur from an action but rather one needs to ensure that the least harm possible occurs. Sometimes avoiding an action that has a reasonable probability of resulting in harm will itself result in harm. Hence, action and non-action both result in harm. Consider the case where a person has necrotising fasciitis in a leg. Amputating the leg is clearly a harm but not amputating is a greater harm because it will result in death. The moderate interpretation of the principle allows the calculation, if possible, of which harm is the most important to avoid. For example, this can be viewed not as balancing benefit with harm but deciding between harms. Of course, this is messy because harm and benefit are inextricably connected; amputating a leg to save a life is, viewed one way, averting harm but, viewed another way, is a benefit: promoting continued life.

Both versions of the precautionary principle have played an important role in European social debates and in the formation of laws, policies and regulations; it has figured prominently in the GM crop debates and decisions. This is not surprising given that it arose in the environmental movement in Germany and other parts of Europe during the 1970s. For the most part, its interpretation was moderate as suggested by the German designation *Vorsorgeprinzip* (foresight planning). It was a two-pronged principle. One prong mandated the complete avoidance of some harms. These are harms that are so severe that nothing would make taking such actions rational – even if the probability of harm occurring were low. Nuclear war would be a prime example. Certain environmental catastrophes were on the list of harms that are imperative to avoid (e.g. loss of biodiversity, continued deforestation, production of greenhouse gases). The other prong involved risk assessment and risk management. The harms dealt with by this prong were not catastrophic but were risks that resulted from actions accompanied by the reduction of another harm or the achievement of a benefit.

Hence, at its inception, the principle encompassed risk assessment and management for some harms, and prohibition for others. Since the 1970s, social and political debate has centred on which harms fall into which category. Advocacy groups that were concerned about a particular harm or class of harms sought to convince people that that harm or class of harms should be on the prohibited list. Opponents sought to have it on the risk management list. Some advocacy groups placed so many harms on their prohibited list that their views constituted a rejection of the moderate version and an embracing of the strong version.

An example of the use of the principle can be found in diethylstilbestrol (DES) debates in Europe in the late 1970s and early 1980s. DES is an anabolic steroid, oestrogen. The synthesis of DES was first reported by Edward Dodds and his co-workers in *Nature* in 1938 (Dodds *et al.*, 1938). The US Food and Drug Administration approved its prescription use in 1941. An early 'off-label' use (i.e. a use which is not on the list of uses for which it was approved[4]) was the prevention of miscarriages. After six years of use for the prevention of miscarriage, that application was added, in 1947, to the list of approved uses. Like many pharmaceuticals prior to the 1970s, DES was deemed safe for use during pregnancy; thalidomide, used for morning sickness in pregnancy, is another dramatic and tragic example of a pharmaceutical deemed safe during pregnancy. As a result of the negative experience with DES, thalidomide and some other pharmaceuticals, the current presumption is that no pharmaceutical is safe unless there is incontrovertible evidence to the contrary. Regrettably, after hundreds of thousands of women had been prescribed DES, in 1953 William Dieckmann and his colleagues published the results of their double-blind study (Dieckmann *et al.*, 1953) (a double-blind study is one in which neither the subjects nor the researchers know which group is receiving the intervention); miscarriage rates were the same for those who received DES and those who did not. The conclusion was that DES is not efficacious for that use. As happens with an alarmingly high frequency – even today – medical practice lagged significantly behind the research. Notwithstanding

[4] Once a pharmaceutical has received approval (in Anglo-American countries), physicians can prescribe it for any use for which they deem it effective subject only to the standards of defensible medical practice. That standard would not be met if it were prescribed for a condition after compelling evidence was available that it was ineffective in treating that condition and potentially harmful. Potential harm alone would be insufficient since if it is effective, risking the harm for a known benefit might be rational.

this study, DES continued to be promoted for this use and physicians continued to prescribe it well into the 1960s. Sales did decline after the publication of the study, so it did have some immediate impact on clinical practice but the decline over the next decade was slow.

By the late 1960s, evidence was emerging that DES was connected to development of clear cell adenocarcinoma of the vagina in the daughters of mothers who had used DES; other disorders connected to DES also began to emerge:

> Rarely, clear cell adenocarcinoma of the vagina occurs in adolescent girls whose mothers used diethylstilbestrol (DES), a synthetic nonsteroidal estrogen, during pregnancy. The DES effect is the first implication of transplacental carcinogenisis in humans [cancer caused by a compound crossing from the mother's bloodstream to the fetus through the placenta (an organ attached to the wall of the uterus connecting the fetus and the mother and shed during birth – sometimes called 'the afterbirth')]. In females exposed to DES, the following abnormalities have been observed: abdominal preovulatory mucus, a T-shaped endometrial cavity [the endometrium is the lining of the uterine cavity; 'endometrial' is the adjective], menstrual dysfunction, spontaneous abortion, incompetent cervix, and increased incidence of ectopic pregnancy and preterm labor [in an ectopic pregnancy, the placenta attaches outside the uterus – usually in a Fallopian tube (the tube through which an ovum – egg – travels from an ovary to the uterus)]. (Beers and Berkow, 1999, p. 2024)

That was the human tale and a tragic one at that. There is another DES drama and one directly connected to the use of the precautionary principle. In 1947, Purdue University researchers discovered that DES acts as a growth hormone in heifers (young cows that have not had a calf). DES began to be used commercially as a growth hormone in heifers (the source of veal) shortly after this discovery. During the 1970s many groups lobbied for a ban on the agricultural use of DES without success. Industry interests trumped consumer group concern. Consumer groups were unable to match the political power and funding of the beef industry; 1980, however, marks a turning point. Alarmist media coverage heightened public concern. Both Peterson and Caduff cite the so-called 'Italian infant scandal' as an example:

> Allegedly, cases of babies menstruating and growing breasts had occurred in Italy. While the allegations were never proven . . . (Peterson, 1989, p. 461)

> After Italian magazines reported that DES-enriched veal in baby food had led to abnormally large genitals and the onset of menstruation among young children, many consumers began boycotting veal products on their own. (*Die Zeit*, 1980/No. 44 in Caduff, 2002, p. 7)

This appears to have been triggered by the discovery of high levels of oestrogen in veal and the imposition of a court-ordered ban:

> ROME (UPI) – State police and health officials all over Italy yesterday confiscated veal suspected of being adulterated by use of estogens. Favorite Italian dishes based on veal vanished suddenly from restaurant menus. Butchers and importers of veal raised angry protest about the millions of dollars the sudden edict was going to cost them. The ban on veal resulted from an order issued by a magistrate in Latina, near Rome. It was supported by the health ministry and special squads of carabinieri (state police) started confiscating supplies in all major cities. The magistrate acted after heavy concentrations of estrogens were found in canned meats and other products sold by a major processing firm near Rome. (UPI article as carried in *The Montreal Gazette* – 25 September 1980)

A little fear mongering goes a long way towards galvanising public opinion and in 1981 the EC (European Community – later to become the European Union (EU)) banned its use. As Caduff notes, this is a clear example of consumer interests prevailing against industry *and scientific knowledge*:

> The analysis of EU regulatory activity illuminates how organized consumer interests can achieve decisive political influence on the basis of public pressure and privileged institutional access, in this case via the European Parliament. European consumer interest groups were thus successful in pushing EU bodies towards more stringent regulations, even though scientific evidence for health risks associated with the use of growth hormones was "thin" at best. Industry interests opposing restrictive regulations were unable to shape the regulatory outcome because of problems in gaining institutional access, and because they were unable to significantly influence the wider public's risk perceptions in this area. In the end, a broad and stable coalition supporting a growth hormone ban emerged, comprising consumer interest groups, national and supranational (EU) regulators, and some producers (primarily agricultural ones). (Caduff, 2002, p. 3 – Caduff, for simplicity, makes no distinction between EC and EU; see her footnote, p. 2)

This partial ban was followed, in 1985, by a complete ban on the use of all growth hormones, naturally occurring and synthetic, even though the EC's own Lamming Committee concluded, in 1982, that use of the three natural growth hormones in meat production did not pose a significant human health risk. A phrase used in the EC Directive 81/602, which banned all use of growth hormone in livestock, invokes the precautionary principle: 'their safety has not been *conclusively* proven' (emphasis added). This is the very high hurdle of the strong version of the precautionary principle; unless it can be *conclusively* proven that no harm will occur, an action should not be undertaken. As noted already, this effectively stifles all new activities – and many existing ones – since almost nothing we do is without risk of harm. Julian Morris captures well the advocacy group dynamic at work here:

> Consumer groups, seeing an issue on which they might garner public support, then began demanding a ban on the use of hormones in all livestock production. In 1981, the EC banned the use of DES and established a body of scientists to look into the effects of five other hormones that are commonly used as growth promoters. The body, known as the Lamming Commission after its head, Professor G. E. Lamming, issued an interim report in 1982 which concluded that the three natural hormones (estradiol, progesterone and testosterone) 'would not present any harmful effects to the health of consumers when used under appropriate conditions as growth promoters in farm animals'. Indeed, it will not usually be possible to identify meat from cattle treated with these natural hormones as residual levels are typically within normal variability observed in untreated cattle. However, in 1985, before the Lamming Commission had completed its research into the effects of the synthetic hormones, the European Commission – under increasing pressure from consumer groups – banned the use of all growth hormones, with effect from January 1988. (Morris, 2000, p. 2; Morris cites World Trade Organization, 1998, and *The Cargill Bulletin*, 1999 – the latter is an online resource and seems no longer to be available)

This is an example of the strong version of the precautionary principle being used to circumvent any risk/benefit analysis. The outright ban of all growth hormone use (artificial and synthetic) in livestock was clearly not based on the best scientific knowledge but on political and trade interests. Hence, some justificatory principle is needed, one that renders otiose the assessment of risk based on evidence and analysis. The strong version of the precautionary principle was ideal. It justified the action because 'their [the hormones] safety has not been conclusively proven', and the principle requires conclusive evidence that **no** harm will occur for their use. It also has the additional benefit of portraying decision-makers as vigorously protecting the health and safety of consumers.

This strong version of the principle held sway for only a brief period. Whatever its utility in the specific context of the growth hormone controversy, as an entrenched principle it is unsustainable. Since it is impossible, within science and technology, to provide *conclusive* proof that the use of new (and, for that matter, old) products and technologies is safe (free from harm), a consistent appeal to the strong version of the precautionary principle, effectively, would require eschewing science and technology – to the limit, a return to the 1700s or earlier. There may be some people who would support such a position but most would not. Consider, for example, clinical medicine. This is an arena of science and technology where risks abound – a fact to which we shall return later. Clinical medicine would wither were the strong precautionary principle to be applied to it. The attractiveness of the strong version of the precautionary principle lies in its piecemeal employment. It may help an advocacy group or a government to 'sell' a position, or justify a decision, on this or that issue, but its consistent application to all decisions and positions is untenable. Hence, it fails the analytical test of consistency.

Having embraced the principle in the growth hormone case and a few others, the EU and its member countries soon found that the principle had become its sword of Damocles.[5] It was on the horns of a dilemma. On one horn, consistent application of the principle was clearly untenable, and obviously not in any governing body's interest since it constrains the deliberative and decision-making role of such bodies. On the other horn, since the EU had employed and espoused the principle, thereby tacitly accepting it, to repudiate it now would seem erratic and also signal a softening of its commitment to the protection of the health and safety of the public. The task, when on the horns of a dilemma is to find a middle way between the horns. That is what

[5] Damocles, in Greek legend, was enamoured with the power of Dionysius, ruler of Syracuse, and constantly exclaimed how fortunate Dionysius was. In response, Dionysius held a feast where Damocles was wined and dined like a ruler but he had a sword placed over his head hanging by a single horsehair. This was designed to illustrate how precarious is the life of a ruler. By endorsing the precautionary principle, the EU had placed a sword of Damocles above its own head.

the EU did in 2000; its Commission issued a communication on the precautionary principle: COM(2000) 1. The communication accepted the principle but provided a definition that significantly diminished its potency, thereby sliding off both horns into the middle ground.

Point 2 in the summary of the communication reads:

2. The Commission's fourfold aim is to:
   - outline the Commission's approach to using the precautionary principle,
   - establish Commission guidelines for applying it,
   - build a common understanding of how to assess, appraise, manage and communicate risks that science is not yet able to fully evaluate, and
   - avoid unwarranted recourse to the precautionary principle, as a disguised form of protectionism.

Note the allusion to the fact that the precautionary principle has been used 'as a disguised form of protectionism'.

Two foundational elements of this clarification of the scope and application of the principle are found in points 4 and 5 of the summary:

4. The precautionary principle should be considered within a structured approach to the analysis of risk which comprises three elements: risk assessment, risk management, risk communication. The precautionary principle is particularly relevant to risk management.
5. Decision-makers need to be aware of the degree of uncertainty attached to the results of the evaluation of the available scientific information. Judging what is an "acceptable" level of risk for society is an eminently *political* responsibility. Decision-makers faced with an unacceptable risk, scientific uncertainty and public concerns have a duty to find answers. Therefore, all these factors have to be taken into consideration.

Factors that guide decision and action are described in point 6:

6. Where action is deemed necessary, measures based on the precautionary principle should be, *inter alia*:
   - *proportional* to the chosen level of protection,
   - *non-discriminatory* in their application,
   - *consistent* with similar measures already taken,
   - *based on an examination of the potential benefits and costs* of action or lack of action including, where appropriate and feasible, an economic cost/benefit analysis,

- *subject to review*, in light of new scientific evidence, and
- *capable of assigning responsibility for producing the scientific evidence* necessary for a more comprehensive risk analysis.

Banished is the strong version of the principle; affirmed is a commitment to the tenets of risk assessment, management and communication. The principle has been gutted of any meaningful force; it is a phrase, the content of which is entirely based on the principles and methods of risk assessment, management and communication. That is where the EU stands today and it is where almost all rich countries stand.

4 The controversy

Ideological and theological objections

The controversy over agricultural biotechnology has two faces. One face, the one addressed in this chapter, has to do with philosophical, theological or, more broadly, ideological commitments. These include concerns about humans 'creating' new forms of life, about violating or transgressing some fundamental ethical or theological principles, about an economic system in which commodities essential for health – indeed survival – such as food, water and therapeutics are controlled by private enterprises – through patents, concentration of the means of distribution and the like – and the transformation of human existence. The other face, addressed in the next two chapters, has to do with presumed benefits and harms, and balancing them. There are benefit and harm issues embedded in the philosophical, theological (Brunk and Coward, 2009) and ideological concerns, but those typically are not the focus, motivation or underpinnings of the concerns or their resolution.

Much of the controversy explored in this chapter arises from the views of advocacy groups and non-governmental organisations (NGOs): religious denominations, aid organisations, lobby groups and the like. Understanding some facets of these groups is an essential part of coming to terms with their positions, tactics and motivations. Hence, I begin this chapter by glancing into the world of advocacy groups.

4.1 Advocacy and NGOs

A thriving democracy, social engagement and social progress require a marketplace of ideas. A marketplace of ideas requires tolerance and pluralism. Not everyone has a reasonable or prudent idea, but allowing him to express an eccentric view is at the core of a free, open and democratic society. There are, of course, limits; inciting people to harm others is not consistent with a tolerant, pluralistic society and it demeans us all. The limits are captured by the notion of a civil society, not an easy idea to define but essential. There are also tactics that undermine a vibrant marketplace of ideas. One of special importance when examining advocacy groups and their tactics is the casting of issues and claims in terms of 'black or white', 'right or wrong', 'truth or lies', 'villains and saviours', and 'for us or against us'. This strategy, which is very successful in recruiting converts, removes any middle ground even though the middle ground is most often the more rational, prudent and socially desirable place to be.

This language of 'for us or against us' often signals a clash of rigid ideologies. Hence, a few words about ideology are in order. An ideology is a system of beliefs (including myths and doctrines such as religious or political dogmas) that informs individual or social actions, claims and attitudes. Expressed this way, having an ideology is unavoidable. It would be disingenuous for anyone to pretend that she did not approach an issue with preconceptions; it would also be disingenuous to deny that a currently held system of beliefs plays a role in one's interpretation of experience and information. Problems arise when an ideology is rigid: not open to question or revision. It is taken as the light and truth. Even more problematic is a rigid ideology that regards all other ideologies as fundamentally dangerous (forces of darkness) and, hence, not to be tolerated. Rational analysis and decision-making do not depend on a pretence to being an empty vessel waiting to be filled or a blank slate waiting to be written on; this asks the impossible. What is required is a willingness to examine different perspectives, to entertain challenging questions, to pursue answers and to hold views tentatively, expecting that they will change many times during one's life as a result of new information and analytical examination. Those having rigid ideologies are not like that. A person with a rigid ideology will know that a particular scientific finding is false or a person's motives are impure because his ideology entails that that is the case. The task at hand for the rigid ideologue is to ensure, using whatever techniques are available, that the false view is quashed.

It has become common to describe rigid ideologues as dogmatic. That does not quite capture this stance but comes close. So, in what follows, 'dogmas' will describe rigid ideologies and 'dogmatic' will describe rigid ideologues.

Ideally, statements individuals, groups and institutions make should be reliable. Regrettably, this all too often is not the case. Depressingly, even governments (governments in democratic countries) play fast and loose with the truth. After the fact, for example, statements by EU countries that pointed to concerns about product safety as a justification for non-approval and importation bans on products such as GM crops turned out to have been motivated mostly by concerns about trade. This was because imposition of a protectionist trade barrier is immediately open to challenge through the World Trade Organization in a way that a claimed safety concern is not; hence, decisions supposedly based on safety concerns are less open to an international remedy. These issues usually get sorted out internationally over time; the interdependence in numerous spheres is too substantial. Corporations are also ingenious at deception but again there are regulatory frameworks and legal frameworks that ultimately ensure product, process and marketing claims can be justified. It is both disturbing and reassuring that, on 2 September 2009, US Department of Justice lawyers announced a $2.3 billion out-of-court settlement with Pfizer. This was probably the largest settlement of a health-care fraud lawsuit in US history. It is depressing, because Pfizer, a pharmaceutical giant, was tacitly admitting some level of guilt in a serious fraud case. This kind of fraud in a health-care arena – an arena in which we all place considerable confidence – does not bode well for confidence in the integrity of large, multinational corporations. On the other hand, it is reassuring because the legal and regulatory systems worked to expose the fraud, penalise the corporation and redress the harm.

Dogmatism and playing fast and loose with the truth are widespread in advocacy groups and governments (in the quest for electoral success, governments have come to behave more and more like advocacy groups). Clearly, this is not true of all advocacy groups, and the degree varies considerably among those that do succumb to these features; this, as I have suggested is true of almost every issue, is not black and white. In addition, they often emerge over time as an advocacy group crystallising its message and struggling for funding and a continued existence. The dominant kind of advocacy group in rich countries is the NGO. This non-governmental, not-for-profit sector whose existence I strongly support, and with which I have been involved, causes me greater concern than either governments or corporations. They are much more loosely regulated and monitored and are given considerable latitude in the views they express, the 'evidence' they adduce, the causes they champion, their communications with donors and their financial accountability. Since I support the existence of this sector, before I articulate my concerns, let me focus on the positive features – the features that motivate my strong support. Let me also be clear, although I imagine it will be obvious, that, although shared by a significant number of individuals, this is a personal perspective and somewhat editorial in nature. First, and undeniably, most NGOs raise the level of political awareness and achieve humanitarian goals that would otherwise remain unaddressed. At times, one might wonder why much more has not been accomplished – why, for example, poverty in sub-Saharan Africa seems to persist and in some regions to be increasing – but the explanation is multifaceted and NGOs are only one part of a complete explanation. The success of NGOs has mostly been in humanitarian relief – providing medical care, access to clean water and education. Their activities have been less successful in ameliorating systemic problems faced by low- and middle-income countries – problems such as government corruption and inefficiency, aid dependence, trade subservience and the like. Indeed, along with rich nations' government aid programmes, it is arguable that the humanitarian relief has frustrated the tackling of systemic problems. I return to this in several places below.

Second, notwithstanding all the rhetoric, governments are obsessed with public perception and are pathologically secretive. The obsession with public perception – something that political parties believe is a requirement of retaining or gaining power – is debilitating. Governments are immobilised by layer upon layer of regulation and bureaucracy, which has grown haphazardly, to convince the broad electorate that there is equal access to programme funding, effective stewardship of resources, no unwarranted benefit to individuals or groups and so on. These are admirable goals. Regrettably, the undesirable behaviour still occurs, as a litany of recent scandals attest. The plethora of complicated rules, regulations, monitoring and approval processes seems to result only in a climate of timidity to act and an attendant slow grinding of the gears of action. NGOs are more nimble and often achieve social goals that governments find difficult to achieve because they have to respond to a cacophony of different and usually incompatible voices. An NGO needs to satisfy a self-selected portion of the population (its donors); it needs to convince them that it is achieving its objectives. This gives their activities focus and reduces the cacophony of different voices found in the public arena.

The need to pay attention to this cacophony of different voices is one reason governments are exceptionally prone to secrecy. The less people know, the fewer the things for them to criticise, or to agitate for or against. There obviously are legitimate reasons for secrecy. Sometimes national security is at stake but political secrecy is far too pervasive for that alone to be a credible explanation. In this environment, NGOs play another useful role; a subset of

NGOs serve as public watchdogs. They have the resources, the trained staff and a committed audience of supporters, and can garner media attention to issues. Their independence from the political parties allows them to explore and expose issues and information that political parties – especially ones in power – wish to suppress.

Third, NGOs enable individuals to participate in society in a meaningful way. The plethora of NGOs with rich and varied goals allows individuals to support (financially or through participation) social objectives with which they identify. Furthermore, they allow citizens to identify and pursue social objectives with which those holding the reins of political power may not agree.

These features of NGOs enrich participatory democracy. A free and fair election every few years is mistakenly seen as the hallmark of democracy; for sure, it is an important component and a precondition for a democratic state, but the vibrancy of a democracy rests with the facilitation of a free and open exchange of ideas (a marketplace of discourse on ideas), and an independent judiciary able to protect enshrined rights and freedoms. NGOs and the media are a large part of the facilitation of a free and open exchange of ideas. An independent judiciary is able to protect enshrined rights and freedoms, and provide recourse for those who believe legislators have encroached on those rights and freedoms. To exploit a question posed by Socrates in Plato's dialogue *Euthyphro*:

> **Socrates**: And what do you say of piety, Euthyphro: is not piety, according to your definition, loved by all the gods?
> **Euthyphro:** Yes
> **Socrates**: Because it is pious or holy, or for some other reason?
> **Euthyphro:** No, that is the reason.
> **Socrates**: It is loved because it is holy, not holy because it is loved?
> **Euthyphro:** Yes.
>
> (Jowett, 1931, p. 13)

My co-opting of this distinction runs, 'Is something right because the legislators of the day deem it is right or do they deem it right because it is right?' The history of Western social and legal evolution has come to weigh heavily against 'it is right because legislators deem it right', this being too open to the vagaries of ever changing political whims, fashions and dogmas – not to mention malevolent, oppressive regimes such as Hitler's. This case is well stated by John Stuart Mill, 'The aim, therefore, of patriots was to set limits to the power which the ruler should be suffered to exercise over the community; and this limitation was what they meant by liberty. It was attempted in two ways. First, by obtaining a recognition of certain immunities, called political liberties or rights, which it was to be regarded as a breach of duty in the ruler to infringe, and which if he did infringe, specific resistance, or general rebellion, was held to be justifiable. A second, and generally a later expedient, was the establishment of constitutional checks, by which the consent of the community, or of a body of some sort, supposed to represent its interests, was made a necessary condition to some of the more important acts of the governing power' (Mill, 1859, pp. 8–9). Instead most democratic societies have demanded that legislators should deem 'actions, laws and regulations' right because the citizenry has deemed them right, and enshrined it as right. A charter of rights and freedoms, or a bill of rights, sets out that which a particular society has enshrined; an independent judiciary protects it from legislative disregard. If there is social agreement, a charter or bill of rights might be able to be amended, but until an amendment occurs, legislators, like all citizens, are required to respect it. The activities of many NGOs complement the judiciary in protecting entrenched rights and freedoms. In some cases, NGOs might even hold the judiciary itself to public account in this respect.

So that is the positive case, the reason that the existence of NGOs is essential to a vibrant democracy. Now, my concerns get voiced. The first concern arises from the fact that large NGOs have executives, office staff and fieldworkers; they have office buildings, transportation costs and so on. Employees of NGOs depend on it for their incomes. Buildings need to be maintained, heated and cooled, and cleaned, and often mortgages or rent have to be covered. As a result, there is a very strong and ubiquitous incentive to ensure the continued existence of the NGO. This, in turn, results in some clear distortions. Were, for example, an NGO to achieve all its goals, which is after all the point of an NGO, the *raison d'être* for its existence would evaporate. Why would donors contribute to a goal that has been achieved? For example, an NGO that arose to combat nuclear weapons proliferation would no longer be needed if a comprehensive international non-nuclear-proliferation treaty were ratified. Faced with this, an NGO might simply disband, but the above-cited incentives for continuing it make that an unlikely outcome. More likely the NGO will look for a new 'cause' (or causes) to ensure donations continue. Once it is large, with thousands of people's futures and livelihoods at stake, disbanding after achieving major objectives is hardly ever an option. Fully achieving the goals of an NGO is the extreme case. However, even if the goal has not been fully met, donor fatigue with the objective, diminishing returns for effort, etc., may require a shift in focus.

New objectives that do not appear to be an obvious departure from the original motivations of the NGO or that have not already been taken up by others are not easy to find. Hence, one can understand the temptation to manufacture an objective, to find an issue that can be moulded into a cause. One technique of moulding a cause to induce donors to keep giving is to exploit, and indeed foster, public distrust of science, technology, governments, corporations and so on, as well as exploiting public fear of the unknown and public ignorance of science. This is all too frequently adopted as a strategy and is depressingly destructive.

A second concern is that NGOs that are not-for-profit and are not registered charities experience little government-agency interference in their financial affairs and programme delivery. Things are slightly different for registered charities – the majority of NGOs. They have to meet certain requirements with respect to issuing charitable receipts, spending on programmes and eligible charitable activities. Beyond those, for the most part, charities are also unregulated. Of course, all the normal legal requirements apply to all NGOs, such as the illegality of fraud, false or deceptive advertising, libel and so on. This minimal regulatory oversight provides charities with considerable scope in organising their activities: delivery of programmes, financial accounting, donor management and the like.

Consider the accepted percentage a charity spends on non-programme expenses – everything from soliciting donations to filing tax returns. The American Institute of Philanthropy considers 60 per cent expenditure on programmes to be satisfactory but indicates that the best charities spend 75 per cent or more of donations on programmes. Anything below 60 per cent is unsatisfactory. Donors usually demand that a charity meet the highest standard. Most large, efficient charities report programme spending of 80–85 per cent. This, however, is a slippery arena; there is much room to be creative. Some things (e.g. donor recognition programmes) seem clearly not to be charitable activities; other things (e.g. distributing mosquito nets in rural Kenya) will be accepted by everyone as a charitable activity. What, however, about an end-of-project dinner, with entertainment and an open bar, for fieldworkers after completion of construction of a medical clinic? Opinions on this will differ but some charities will fold this into the food bill entry in the accounts for the project and few people will ever detect it; some others will not. Some donors will not mind, thinking this a legitimate expense to keep the spirits and motivation of fieldworkers high; others will find this a travesty and the subterfuge totally unacceptable. Of course, one celebratory dinner in three months – even if it should ideally have been part of the non-programme expenditures – is unlikely to send shock waves through the donor base, but there is no guarantee that this is a very rare one-off incident. Quite the contrary, given understandable donor intolerance for more than 20 per cent going to non-programme expenditures, and the difficulties of effectively running an operation within that percentage, one can expect that only the absolutely obvious items will be assigned as overhead. A large number of other 'operational' costs will be assigned to projects.

Most charities, like most other organisations, do disclose their financial statements. Often people confuse 'full disclosure' with transparency. In a complex, multifaceted organisation, the financial statements are also complex. Only a rigorous financial audit by accountants with expertise in auditing and forensic accounting can really be expected to drill through the high-level disclosure of public financial statements. That level of independent auditing is required of government departments in most rich countries but not of charities.

In summary, advocacy groups, especially NGOs, have an essential role in free, open and democratic societies. They function as watchdogs; they motivate public action; they contribute to a vibrant marketplace of ideas. Older and larger advocacy groups, however, face many of the same pressures for survival that businesses face. For example, where many businesses have to satisfy shareholders and consumers to remain viable, advocacy groups have to satisfy their donors, and remaining viable for both becomes part of the motivation for decisions and actions. Furthermore, businesses are much more regulated than advocacy groups. Many, after the economic meltdown in the USA and Europe in 2008, may think the regulations on banks, for example, were inadequate or not effectively enforced, but the fact remains that advocacy groups face far fewer restrictions on their activities. These concerns have caused me to be cautious in accepting the claims of advocacy groups; no matter how much I might identify with a noble goal they claim to be striving to achieve, and there are many, I know that their motivations, their tactics and their information are subject to all the same pressures as businesses, of which I also have a healthy scepticism.

4.2 Interfering with life

In his *Introductory Lectures on Psycho-Analysis*, Freud (1979) wrote:

> In the course of centuries the *naïve* self-love of men has had to submit to two major blows at the hands of science. The first was when they learned that the earth was not the centre of the universe but only a tiny fragment of the cosmic system of scarcely imaginable vastness. This is associated with Copernicus . . . The second blow fell when biological science destroyed man's supposedly privileged place in creation and proved his descent from the animal kingdom and his ineradicable animal nature . . . But human megalomania will have suffered its third and most wounding blow from the psychological research of the present time which seeks to prove to the ego that it is not even master in its own house, but must content itself with scanty information of what is going on unconsciously in its mind. (pp. 284–285)

Another blow at the hands of science – perhaps the most decisive – fell a little more than 40 years later; by the 1980s, the full impact of this blow had been felt. Living things including humans are a mere collection of chemicals – a complex collection with self-organising and autocatalytic properties to be sure, but chemicals all the same. Most significantly, we have learned how to manipulate these chemicals to alter forms of life, creating novel living entities.

Humans moved from being shaped by the physical world to shaping that world over many millennia. The harnessing of fire, development of the wheel, domestication of animals and crop agriculture, to mention a few key developments, all inexorably enabled humans to dominate nature rather than being subservient to it. During the last two centuries, the pace of domination has been accelerating dramatically. There have always been costs of this domination; for example, clearing forests for crops led to erosion, and increased reliance on domesticated animals and cultivated crops for food and clothing risked starvation, exposure and death from the ravages of disease to crops or animals. Also, communal living and the close proximity to animals that domestication required created problems of sanitation and disease (Diamond, 2002).

There is nothing new in the manipulation by humans of the world about us. We have created, for example, synthetic chemicals, modified the structure of materials, and fractionated substances such as crude oil into gasoline (petrol). So, creating novel entities in the physico-chemical realm is far from new. There have been harsh lessons along the way and there is no shortage of current challenges from our past and present manipulations of nature. We have also manipulated aspects of the living world. For example, through animal and plant breeding, we have created breeds of dogs and farm animals and crops that would not exist were it not for human artificial selection and hybridisation (many of these cannot survive without human attention). We have artificially changed physiological processes, as with ACE inhibiters, such as ramipril, to lower blood pressure, and oestrogen-progesterone pills to suppress ovulation, and the list could go on and on.

In light of this history, GM, reasonably, might be viewed as just one more step in the familiar journey to manipulate nature to suit our purposes. Not surprisingly, many people challenge this position. There are those that oppose, to a significant extent, reliance on science and technology or even the claims of science. Sometimes opposition is a response to the hubris involved in relying on science and technology or enshrining scientific claims; sometimes opposition reflects a conflict between deeply held views and the claims and theories of science, a classic example of which is the rejection by fundamentalist Christians and Muslims of the fact and theory of evolution, the age of the Earth and the astrophysical origins of the universe. An extreme expression of this opposition, and one that few espouse, is that science and technology constitute an unjustifiable intervention in nature that disrupts the natural order and must be stopped. However, even religious fundamentalists who oppose entire domains of science and technology avail themselves of modern medical interventions into nature. Two ironies emerge from this. First, to be an effective clinical physician, one need not, and a significant number do not, accept evolution; the basic biological science on which most clinical advances rest, however, is thoroughly dependent on an evolutionary perspective. Second, and more relevant to GM plants, many recent advances in clinical practice rest on genetically modifying organisms – frequently, but not only, bacteria.

The fact is that even the most ardent opponents do not personally wish to return to an era prior to science or technology – to primeval times. Vilifying this or that domain of science and technology (GM agriculture, for example) while accepting the benefits in another domain (GM medicine, for example) is both inconsistent and depressingly common.

Unlike those who either reject many science-based knowledge claims or reject the technological application of scientific knowledge, many individuals accept that science and technology have benefited us in many ways, and that, on balance, the benefits outweigh the harms. In addition, they recognise that, lapses notwithstanding, we have learned the lessons of history and, hence, continually monitor new scientific and technological innovations for signs of previously unknown harms and then initiate action to mitigate them. Of course, the potential for unpleasant surprises abounds. Arguably, our greatest failing is not ignorance about specific harms arising from science and technology; it is our individual and/or collective difficulty in acting to mitigate them, especially when to do so has immediate and negative effects – especially financial/economic effects.

Reduction of human-produced greenhouse gases is a case in point. The economic, financial and lifestyle impacts of such a reduction are large and unpalatable. Dramatically reducing automobile use (trucks and cars) has to be part of a strategy to mitigate greenhouse gas production, but the consequences for jobs, corporate profits and individual lifestyle are profound. The same is true of greenhouse gases produced by agricultural activity, which, as many have pointed out, far exceed those produced by automobile use – more on this in Chapter 5. For many, the cure seems worse than the disease, but even for those who readily accept that the disease must be cured, the therapy seems daunting, in the face of which the will to act wavers and often falters. This, however, is not a failure of science and technology; it is a failure of willingness to act on the part of individuals and governments.

Among those that accept the value of science and technology, there is a spectrum of acceptance. Some people embrace virtually all innovations in science and technology and focus principally on risk identification, assessment and management. Others are more wary and sceptical and embrace these innovations slowly, cautiously and partially. They may be somewhat comfortable with innovations in medicine, for example, but not in agriculture, or in renewable energy, such as solar or wind technology, but not in novel plastics. With respect to molecular biology, this broad group of science and technology acceptors (from champions to wary acceptors) can be divided into two groups. One group accepts that manipulating the molecular basis of life is just another step in scientific and technological development, a step that introduces new challenges – perhaps new risks – but is essentially a continuation of our past history of manipulating nature. The other group holds that manipulating non-living nature is different from manipulating living things.

While both groups focus on risk identification, assessment and management, the first group applies this strategy equally to the molecular manipulation of non-living nature and living things; the second group rejects the acceptability of the molecular manipulation of living things. Those in the first group do recognise that living things have, not to state the obvious, a life of their own. One lesson from instances of adventitious presence is that controlling the course of life forms is nearly impossible. Hence, living things are more challenging to control. With DDT, we were able to partially rectify the disaster we unleashed by banning its use for all but a small number of applications; in the longer term, natural processes inexorably diminished the environmental impacts. A mistake, however, with a new life form may not be undone so easily; indeed, it is conceivable that undoing it might be impossible. Nonetheless, for the first group, this difference between molecular manipulation of non-living and living things is a 'difference of degree' (the degree of risk and the degree of difficulty in managing the risk); it is not a 'difference in kind'. That is, it is not a fundamentally different kind of manipulation of nature; manipulating DNA is not fundamentally different from manipulating other molecules (organic or inorganic). GM simply poses a more challenging risk assessment and risk management environment than other chemical modifications. These are, of course, excellent reasons for increased caution and vigilance, and for placing exceedingly tight restrictions on GM, but not for imposing a complete ban.

The second group perceives a 'difference in kind' between manipulating non-living nature and manipulating the molecular genetics of living things, a difference that entails that we are crossing a boundary (i.e. crossing a conceptual and philosophical or theological boundary); we are engaging in an entirely new category of manipulation. On this view, adding, deleting or substituting a nucleotide sequence in a living thing such that it remains a viable life form but has one or more novel, 'non-natural' traits is fundamentally different from breaking the double-carbon bond in a number of benzene styrene monomers and causing them to join (polymerisation), thereby creating polystyrene. The upshot, for a large number of people who hold this difference-in-kind position, is that genetic modification is more than imprudent; it is morally unacceptable and should be prohibited.

The tenability of this view depends, first and foremost, on identifying some properties of livings things not possessed by non-living things or vice versa. If that case can be made, a second requirement is that, in light of the difference, GM is immoral. In what follows, I argue that neither of the requirements that must be met for this view to be tenable is obviously true; indeed, there are reasons to be suspicious that accepting either of them is reasonable.

First, let's examine whether a property can be found that differentiates manipulating DNA from manipulating other molecules. Self-reproduction is an obvious candidate; some others are self-maintenance, self-organising and embryological development. How compelling is the case based on one or more of these properties? Ilya Prigogine was awarded the Nobel Prize in 1977 for his description of, and work on, 'dissipative structures' (or dissipative systems), a term he coined for systems that are thermodynamically open (they are not isolated and energy flows in and out of them) and are far from thermodynamic equilibrium. A system is in thermodynamic equilibrium when there is no net energy exchange between it and its environment (which includes other systems) or within the system itself. A system far from the equilibrium state has high net energy flow within the system, and between the system and its environment. Dissipative systems, Prigogine demonstrated, are self-organising, autocatalytic systems (self-maintaining and self-regulating of its chemical processes). In his book, *Exploring Complexity: An Introduction* (co-authored with Grégoire Nicolis and designed to be more accessible than his technical books and papers), he characterises the implications of his work on dissipative structures:

> Such mechanisms [open systems in a non-equilibrium state] are known to exist in chemistry, and their most striking manifestation is autocatalysis. For instance, the presence of a product may enhance the rate of its own production. As a matter of fact this seemingly exotic phenomenon happens routinely in any combustion process, thanks to the presence of free radicals, those extremely reactive substances containing one unpaired electron, which by reacting with other molecules give rise to further amounts of free radicals and thus to a self-accelerating process. In addition, *self-reproduction*, one of the most characteristic properties of life, is basically the result of an autocatalytic cycle in which the genetic material is replicated by the intervention of specific proteins, themselves synthesized through the instruction contained in the genetic material. (Nicolis and Prigogine, 1989, pp. 17–18)
>
> ...
>
> Being convinced by now that ordinary physico-chemical systems can show complex behaviour presenting many of the characteristics usually ascribed to life, it is legitimate to enquire whether some of the above features of biological systems can be attributed to transitions induced by nonequilibrium constraints and appropriate destabilizing mechanisms similar to autocatalysis. This is probably the most fundamental question that can be

> raised in science . . . The particular problem on which we focus here is the control of embryonic development. (Nicolis and Prigogine, 1989, p. 32)

An account of embryological development then follows. Self-reproduction, embryological development and the like are exceedingly complex processes and there are many features and processes yet to be adequately described. Since the work of Prigogine, however, each feature or process of biological systems has been shown to have an analogue in physico-chemical systems. As Nicolis and Prigogine (1989) note:

> Living beings are undoubtedly the most complex and organized objects found in nature, in view of their morphology and their functioning. As we have emphasized, they serve as prototypes from which physical sciences can get both motivation and inspiration for understanding complexity. (p. 32)

It is precisely because living systems are not different in kind that they can guide physico-chemical research into complex physico-chemical systems. There are, as one would expect, sceptics of this view. The last 50 years of physico-chemical research and theorising on complex systems and chaotic dynamical systems, however, has continually eroded the evidential and theoretical basis for this scepticism – at least scientifically speaking – and we are here discussing the position of individuals who accept science and technology in other domains.

For many people who claim a difference in kind exists, a scientific difference in kind is not what is meant, so Prigogine's work and scientific discoveries over the last 40 years are not relevant. The difference in kind has to do with matters relating to the meaning of life – to spiritual matters (the question of the purpose of life (especially, my life), for example). Scientific research can contribute to an answer but will not provide a complete or compelling one. This distinction between science and spiritual matters has become commonplace in Western Christian tradition. I concentrate in what follows on Christian religious perspectives on science and spiritual matters because they provide an interesting case study of the development of one form of 'compatibilism', the view that scientific enquiry and discovery are compatible with spiritual enquiry and discovery. Islam might provide an interesting case study of the dissolution of a compatibilist perspective. It appears that during the medieval period, science and mathematics flourished in the Islamic world and were not seen as incompatible with faith; today, that sense of compatibility seems, mostly, to have evaporated.

Christianity – in a process that began at the latest with Galileo – has had to confront the undeniable evidence for the modern neo-Copernican view of the structure of the universe and the Einsteinian theory of its dynamics, and also the now overwhelming evidence for the fact of evolution and its explanation by the modern synthetic theory of evolution. There are, of course, Christian denominations that do not accept this position – most notably fundamentalist evangelical denominations (and sometimes individual congregations) in the USA. Some reject all the scientific claims that conflict with *their* literal reading of the Bible; I emphasise 'their' because no uniformity of interpretation exists among these denominations – indeed, one reason for such a fractured fundamentalist religious landscape is that groups separate over differences of interpretation. Most focus their attention on biological evolution, seeing it as the antithesis of belief in a God who created humans in His own image, endowed them with souls and will reward or punish them in a life after death. Those who believe that the Bible (or the scriptures of any faith) provides knowledge of the natural world and human origins are, in my view, on exceedingly unstable footings. Moreover, this book, accepting as it does current scientific knowledge and theorising, is unlikely to have much relevance to them. So, I focus on the views of more mainstream Christian denominations, to which the vast majority of Christians belong.

The stance taken by most of these Christian denominations is that scientific investigation is the appropriate way (indeed the only way) to discover the nature of things; religious investigation is the appropriate way (indeed the only way) to discover the nature of human existence, the purpose of life and the fulfilment of our spiritual yearnings and destinies.

John Paul II expressed this for the Roman Catholic Church in this way (John Paul II, 1988):

> Both religion and science must preserve their own autonomy and their distinctiveness. Religion is not founded on science nor is science an extension of religion. Each should possess its own principles, its pattern of procedures, its diversities of interpretation and its own conclusions . . . While each can and should support the other as distinct dimensions of a common human culture, neither ought to assume that it forms a necessary premise for the other. (p. 377)
>
> For the truth of the matter is that the church and the scientific community will inevitably interact; their options do not include isolation. Christians will

> inevitably assimilate the prevailing ideas about the world, and today these are deeply shaped by science. The only question is whether they will do this critically or unreflectively, with depth and nuance or with a shallowness that debases the Gospel and leaves us ashamed before history. Scientists, like all human beings, will make decisions upon what ultimately gives meaning and value to their lives and to their work. This they will do well or poorly, with the reflective depth that theological wisdom can help them attain or with an unconsidered absolutizing of their results beyond their reasonable and proper limits. (p. 378)

The Church of England takes a similar approach (Church of England, 2010).

> Questions of science and religion touch the deepest issues of human existence and purpose. Scientists and theologians approach these questions in very different ways. Who cannot be amazed at the beauty, the complexity, the vastness of the created order and wonder at how it came to be? Wonder at the very question of why it exists at all. Or wonder at the fine tuning of the physical constants that allow carbon based life to exist in this order. Or puzzle about how we came to have consciousness and purpose. Or ponder the deep philosophical and religious questions of human existence which, contrary to the views of some well-known atheist scientists, are quite beyond the explanatory power of science and the scientific method.
>
> The rate of scientific development in recent years, which enables us to understand so much more about the world and the universe in which we live, is breathtaking. Within a generation great progress has been made in our understanding of, for example, the nature of the universe, of atomic structure, of DNA and of the genome. These advances have resulted in overwhelming evidence for the truth of many scientific theories, such as the great age of the universe, measured in thousands of millions of years, or its vastness with billions of billions of stars. The discovery of DNA and recent work on genome sequencing is compelling evidence for the interrelatedness of all living things, and the mechanisms of genetic mutation and evolution are now well understood. There is no evidence of any abating in such rapid advances, new discoveries will continue to be made in many areas not least genetics and neuroscience.
>
> For the Christian trying to make sense of this new scientific knowledge, much hinges on how we read the scriptures and how we understand the truth of scripture. There is nothing new about this. When Galileo's observations supported the Copernican theory that the earth and planets orbit the sun this

> was considered to be in conflict with the literal reading of texts such as Psalm 93:1 'The world is firmly established it cannot be moved'. Before the development of modern scientific method and the Enlightenment, questions of whether such a text was literally true in a scientific sense seldom arose. Now we understand that text as absolutely true in a theological and in a poetic sense but not attempting to make a scientific statement. Few today would try to use that text to refute the movement of the planets. Similar questions of interpretation challenge us in other Psalms or in the Genesis accounts of creation, as was noted by Augustine as early as the fifth century. Some will want to read these in a literal way but if we attempt to read scripture as a literal scientific account then inevitably conflict with science results. We do not have to read it that way. If we understand it as complementary to scientific understanding we see a truth no less real, no less important, which gives a completely different level of description to the scientific one. How we do that is an ongoing hermeneutic challenge. (GS (General Synod) 1772A)

Other denominations (fundamentalist, biblical – literalist denominations excepted of course) echo this position on the compatibility of science and religion. Since, science finds no difference in kind between living things and non-living things, the underpinnings of the view that manipulations of DNA are different in kind from manipulations of other molecules in nature must follow from religious or spiritual features of living things that distinguish them from non-living things. There are, to understate the case, lots of challenges to this basis for the claimed difference, but let's accept it, at least for the purpose of a thorough examination of the view (this sets things up for a form of *reductio ad absurdum* proof). What are the implications of this view for GM? Does it entail, as some claim, that GM is immoral and should be banned?

I will argue that an outright rejection of GM is an untenable implication. That is not to deny that some people do adopt that position; it is rather that adopting it has some consequences that most people – even most Christians and other religious people – will find discomforting.

A moral position on GM and the implications of that moral position constitute part of an individual's conceptual framework (ideology); the complete conceptual framework will encompass numerous commitments, beliefs and positions. As a whole, those commitments, beliefs and positions must form a consistent set; recall that during the discussion of tools of analysis in Chapter 3, it was noted that an inconsistent set of claims entails that every claim and its negation (even nonsensical claims) can be proved true. Consequently, since the larger canvas of commitments, beliefs and positions must form a consistent set, a particular position can be scrutinised by assessing its compatibility with other commitments, beliefs and positions. Let's examine in more detail this process of scrutinising from within, so to speak, the commitments, beliefs and positions that make up a conceptual framework.

Suppose a case that was *prima facie* compelling has been made that GM crops were the only viable solution to starvation in low- and middle-income nations. Opponents of GM would face a dilemma; either accept GM crop science and technology or consign millions of people to death by starvation. A committed opponent who saw no other resolution of the dilemma might take the view that GM science and technology is spiritually (or theologically) unacceptable (immoral); hence, if the starvation of millions of people is unavoidable without GM science and technology, so be it. That, for many whose opposition is spiritually and morally based, will be a disturbing path to take. For most religious people who are opposed to GM crops, the dilemma will not be so easily resolved because they will find consigning millions of people to death by starvation as immoral as GM crop science and technology. Consequently, an inconsistency exists in their conceptual framework. Obviously, there are many strategies that can remove the inconsistency. One might prioritise one's moral judgements, so that some moral judgements outrank others. That, most likely, is what a committed opponent of GM would do. The bite-the-bullet position on millions starving is accepting that GM science and technology is more immoral than letting millions of people starve. Large numbers of people will reverse the priority and hold that allowing millions of people to starve is more immoral than employing GM science and technology. These people might restrict GM by requiring that employing GM science and technology is only moral when it is essential to the prevention of mass starvation. As we will see in a moment, this is likely to be a hollow restriction. Another strategy is to eliminate the dilemma by discovering a way (or championing a currently purported way) to prevent starvation that does not employ GM technology. This strategy relies heavily on empirical knowledge and extrapolation. The arguments will not be about morality but about the empirical evidence that a non-GM way of preventing mass starvation is viable/credible. That strategy, like the prioritising moral precepts, faces storms and rocks at every turn in the journey, a journey that I now trace.

In most low- and middle-income countries, population growth has already outstripped the means of subsistence, and individuals live in various states of poverty, starvation, inadequate nutrition and poor health. A word on constantly changing nomenclature is in order here; the terms 'developed countries' and 'developing countries' are now, frequently, replaced by 'rich countries' and 'low- and middle-income countries'. This has the advantage of providing a more precise (quantitative) description of the economic status of countries and their people. Hence, these are the terms I shall use.

Some low- and middle-income countries (e.g. China and India) have made significant advances in ameliorating poverty in some segments of the population. As low- and middle-income countries increase the affluence of their citizens, however, the demand on food supplies increases. Even modest increases in affluence in these countries allow some portion of their citizens to expand the quantity and variety of food consumed. In effect, as the poor become more affluent, the effective population size, with respect to demand on food resources, increases. Although no new individuals are added to the world population, many individuals are added as increased consumers of food. Those who could barely find the resources for a bowl of rice each day will, with a very modest increase in affluence, be able to afford three bowls a day and perhaps a chicken and the like on many days of the year. Even if world population numbers were held constant at 2008 levels (6.7 billion), the demand for food (driven by increasing affluence in low- and middle-income countries) will rapidly outstrip current world production. Of the 6.7 billion people in 2008, about 1 billion were in rich countries. The other 5.7 billion were in low- and middle-income countries. Hence, even assuming no future growth in world population, as a significant portion of this 5.7 billion becomes more affluent there will be, at a minimum, a doubling of demand for food. And, not even the most optimistic demographic projections envisage a stabilisation at 6.7 billion. The Population Reference Bureau projection (2008) has the world population at 9.3 billion by 2050 with virtually all of that increase occurring in low- and middle-income countries.

The examinations in Chapters 5 and 7 underscore that meeting this challenge with conventional non-GM agriculture or organic agriculture is impossible unless we want to clear all the forests of the world, devote all non-shelter space to agriculture and continue conventional agricultural practices, which use pesticides, herbicides, synthetic fertilisers and so on. This 'unless', almost everyone would accept, is imprudent in the extreme and some would go farther and declare it a travesty and immoral. Hence, meeting the challenge without GM agriculture is impossible. To glance ahead, three factors contribute to this near impossibility. First, the environmental impact of current conventional farming is very large and is not sustainable. Second, meeting the challenge of future demand for food by non-GM conventional farming – and more so for organic farming – will require bringing more land into agricultural production. That is not a viable path. Third, a significant portion of current farmland is of marginal quality and will continue to require substantial inputs of fertiliser and water unless crops can be engineered to require less of both and still thrive.

Much of this section has focused on those that accept, for the most part, that science and technology have brought benefits that need to be weighed against attendant harms, and that science and technology are compatible with spiritual commitments. A sizable subgroup, however, perceives a clear demarcation between manipulating DNA and manipulating other molecules. For the purposes of analysis and gaining traction of the central issues, the validity of the demarcation was accepted, although I personally do not accept it. The analysis undertaken suggests that the demarcation does not necessarily entail that engaging in GM is always the morally wrong course of action. In fact, individuals who accept the demarcation on spiritual grounds face the same difficult moral dilemmas and decisions (trade-offs) as those who reject the demarcation. Those accepting the demarcation will find themselves, when considering employment of GM, ranking moral precepts; some will trump others; a particular action will be 'the lesser of two evils'. Although in some contexts it is wrong to engage in it, in other contexts it is more wrong not to engage in it. Hence, in those contexts, it is the morally right thing to do after all. The central question is, to what can one appeal to justify a contextual ranking?

If the grounding is based on a weighing of benefit and harm or less harm from doing than not doing, the force of the demarcation dissolves. This is a risk assessment exercise, the same risk assessment exercise that will take place for non-living technological interventions. The same is true if it is based on beneficence or a social contract or a principle of fairness or justice: the force of the demarcation dissolves, since those will also be the underpinning for interventions in the non-living world. Hence, the spiritually grounded demarcation, even accepting its validity, adds nothing; indeed, it is a distraction from the hard task at hand. Furthermore, this suggests that even natural law ethical theorists cannot avoid ranking and some allusion to benefits and harms.

4.3 Patenting life

Advances in science and technology, along with increasingly progressive political and legal changes, have been the principal and essential causes of the improvement of individual and social well-being over the last few centuries. Improvements in health, quality of life (e.g. increased leisure time) and an inexpensive, secure and safe supply of food and water, among others, are a direct result of advances in our scientific knowledge and its technological application. As already conceded, it has been a bumpy ride and has given rise to numerous challenges, some of which have posed, or are posing, severe risks. What is incontestable is that these features of rich nations are, in large part, a result of advances in science and technology. A rich country is not a Utopia but only a little reflection is required to motivate a preference for living in a rich country rather than in a poor country today or at an earlier time. The occasional nostalgia for the simple life of an earlier time is easily banished by reflection on the overall conditions of life in those times – even for the nobility. Life was short, malnutrition widespread and disease common with amelioration unavailable and hard physical labour essential; the life of the nobility was only a little less desperate in these respects. Wind the clock back even 150 years in rich nations and, for most people, life was fraught with dangers; this was a time before vaccines, antibiotics and anaesthetics, to mention only a few things from only one domain: medicine.[1]

The cornerstone of these advances is discovery, innovation and invention; fostering these is, therefore, essential to advancement in science and technology. Legislatures in rich counties have recognised two requirements for fostering discovery, innovation and invention. An obvious one is reward. Sometimes, an individual finds sufficient personal reward in pursuing some research or technological application that no additional reward (such as financial gain) is required. Sometimes, also, the environment in which a person conducts her work provides a reasonable financial as well as personal reward; universities, for example, provide a salary and government-funding agencies provide research support. These are an important part of the overall fostering of

[1] There are reasons to accept that the hyped rhetoric about medical advances is just that – hype. Many have, convincingly, I think, argued that the major gains in medicine came from public health measures (sanitation and water purification), reduction in infectious diseases (through, for example, vaccines) and treatment of bacterial infections with antibiotics (see Le Fanu, 2000; see also Illich, 1977; Kennedy, 1981; Starr, 1982).

discovery, innovation and invention, but are not sufficient. Public funding is usually inadequate or unavailable for essential elements such as industrial-scale production, marketing and distribution; it is often inadequate or unavailable for the initial discovery, innovation and invention since these can be exceptionally expensive (e.g. the development of a pharmaceutical). Private enterprise is a critical component. Private enterprise, however, needs a reasonable return on investment – reasonable includes a guarantee of some profit. If there is no guarantee, other investments – even government bonds – will appear more attractive. There must be a reward for taking a chance on a product or pursuing an invention. A second requirement, connected to but distinguishable from reward, is incentive to disclose. A discovery, innovation or invention that is not disclosed impedes the progress of science and technology.

An array of mechanisms have been put in place to meet these requirements: for example, design rights protecting aesthetic creations, trademarks, copyright, patents and a host of field-specific protections such as, in the agricultural context, plant breeders' rights. They are based on the assumption that protection, for some reasonable period of time, from unauthorised use of an individual's or company's discovery, innovation or invention by others will: (1) permit an appropriate financial reward to accrue to the individual or company, and (2) remove any impediment to disclosure. The name most frequently used to describe the things for which protection is sought is 'intellectual property' (IP). One form of legally entrenched IP protection is dissimilar in many respects to those just listed: confidential information (trade secrets). Trade secrets, as the name indicates, is based on keeping the IP secret. As a result, the IP is never publicly disclosed or 'registered'. Protection is contractual; disclosure to those who are deemed to need to know for some purpose – including an individual or company that uses the IP under licence – is accompanied by a contractual agreement not to disclose the IP. That contractual obligation is enforceable in law. This is an important IP protection in the food industry, the trade secret of the formula for Coca-Cola being a well-worn example.

Biotechnology-related IP is not suitable for protection as a trade secret because so much of the biological knowledge on which the technology is based is available in the public domain. Once a product is announced, discovering its molecular structure and the process by which it was created can be expected to occur quickly. That is why patent protection is the protection of choice in biotechnology. Once disclosed to a patent office and registered, the use of the IP is protected for the life of the patent; knowledge by others does not matter since they cannot use that knowledge in a way that diminishes the patent holder's financial return. The knowledge, however, can be used in further research and the development of new IP. That is one of the goals of providing legal IP protection. Disclosure under a patent makes the knowledge publicly available while protecting the financial and other interests of the patent holder. Although patent protection is the protection of choice in biotechnology, the legal path to obtaining the right to patent biotechnological IP has been tortuous.

Tracing that legal path begins with the definitions and requirements contained in relevant patent acts. Different countries have different patent acts with different provisions and wording. A cursory comparison of the US Patent Act, the Canadian Patent Act and the European Patent Convention will highlight the similarities and differences and provide a context for understanding this tortuous path to patenting biotechnological IP. The relevant sections of each are those that specify what is patentable and what must be disclosed. Although the excerpted sections contain nuanced legal language, the thrust of each section is not difficult to discern.

The US Patent Act in Section 101 defines inventions which are patentable as follows: 'Whoever invents or discovers any new and useful process, machine, manufacture, or composition of matter, or any new and useful improvement thereof, may obtain a patent therefor, subject to the conditions and requirements of this title.' The Canadian Patent Act defines invention as follows: '"invention" means any new and useful art, process, machine, manufacture or composition of matter, or any new and useful improvement in any art, process, machine, manufacture or composition of matter'. It also has a clause indicating what is not patentable (27 (8)): 'No patent shall be granted for any mere scientific principle or abstract theorem.' The European Patent Convention is more expansive on these matters:

> Article 52 Patentable inventions
>
> (1) European patents shall be granted for any inventions which are susceptible of industrial application, which are new and which involve an inventive step.
>
> (2) The following in particular shall not be regarded as inventions within the meaning of paragraph 1
>
> (a) discoveries, scientific theories and mathematical methods;
> (b) aesthetic creations;

(c) schemes, rules and methods for performing mental acts, playing games or doing business, and programs for computers;
(d) presentations of information.

(3) The provisions of paragraph 2 shall exclude patentability of the subject-matter or activities referred to in that provision only to the extent to which a European patent application or European patent relates to such subject-matter or activities as such.

(4) Methods for treatment of the human or animal body by surgery or therapy and diagnostic methods practised on the human or animal body shall not be regarded as inventions which are susceptible of industrial application within the meaning of paragraph 1. This provision shall not apply to products, in particular substances or compositions, for use in any of these methods.

Article 53 Exceptions to patentability

European patents shall not be granted in respect of:

(a) inventions the publication or exploitation of which would be contrary to "order public" or morality, provided that the exploitation shall not be deemed to be so contrary merely because it is prohibited by law or regulation in some or all of the Contracting States;
(b) plant or animal varieties or essentially biological processes for the production of plants or animals; this provision does not apply to microbiological processes or the products thereof.

The Canadian Patent Act has a requirement on specification, a requirement that was pivotal in a Supreme Court decision on patentability of a hybrid plant. It states:

Specification

(3) The specification of an invention must

(a) correctly and fully describe the invention and its operation or use as contemplated by the inventor;
(b) set out clearly the various steps in a process, or the method of constructing, making, compounding or using a machine, manufacture or composition of matter, in such full, clear, concise and exact terms as to enable any person skilled in the art or science to which it pertains, or with which it is most closely connected, to make, construct, compound or use it;
(c) in the case of a machine, explain the principle of the machine and the best mode in which the inventor has contemplated the application of that principle; and

(d) in the case of a process, explain the necessary sequence, if any, of the various steps, so as to distinguish the invention from other inventions.

Claims

(4) The specification must end with a claim or claims defining distinctly and in explicit terms the subject-matter of the invention for which an exclusive privilege or property is claimed.

The US Patent Act has a somewhat similar specification requirement:

Sect. 112. Specification

The specification shall contain a written description of the invention, and of the manner and process of making and using it, in such full, clear, concise, and exact terms as to enable any person skilled in the art to which it pertains, or with which it is most nearly connected, to make and use the same, and shall set forth the best mode contemplated by the inventor of carrying out his invention.

The specification shall conclude with one or more claims particularly pointing out and distinctly claiming the subject matter which the applicant regards as his invention.

A claim may be written in independent or, if the nature of the case admits, in dependent or multiple dependent form.

Subject to the following paragraph, a claim in dependent form shall contain a reference to a claim previously set forth and then specify a further limitation of the subject matter claimed.

A claim in dependent form shall be construed to incorporate by reference all the limitations of the claim to which it refers.

A claim in multiple dependent form shall contain a reference, in the alternative only, to more than one claim previously set forth and then specify a further limitation of the subject matter claimed. A multiple dependent claim shall not serve as a basis for any other multiple dependent claim. A multiple claim shall be construed to incorporate by reference all the limitations of the particular claim in relation to which it is being considered.

An element in a claim for a combination may be expressed as a means or step for performing a specified function without the recital of structure, material, or acts in support thereof, and such claim shall be construed to cover the corresponding structure, material, or acts described in the specification and equivalents thereof.

The US Patent Act specifies conditions for patentability as follows:

> Sect. 102. Conditions for patentability; novelty and loss of right to patent
>
> A person shall be entitled to a patent unless–
>
> (a) the invention was known or used by others in this country, or patented or described in a printed publication in this or a foreign country, before the invention thereof by the applicant for patent, or
> (b) the invention was patented or described in a printed publication in this or a foreign country or in public use or on sale in this country, more than one year prior to the date of the application for patent in the United States, or
> (c) he has abandoned the invention, or
> (d) the invention was first patented or caused to be patented, or was the subject of an inventor's certificate, by the applicant or his legal representatives or assigns in a foreign country prior to the date of the application for patent in this country on an application for patent or inventor's certificate filed more than twelve months before the filing of the application in the United States, or
> (e) the invention was described in a patent granted on an application for patent by another filed in the United States before the invention thereof by the applicant for patent, or on an international application by another who has fulfilled the requirements of paragraphs (1), (2), and (4) of section 371 (c) of this title [35 USCS Sect. 371(c) (1), (2), (4)] before the invention thereof by the applicant for patent, or
> (f) he did not himself invent the subject matter sought to be patented, or
> (g) before the applicant's invention thereof the invention was made in this country by another who had not abandoned, suppressed, or concealed it. In determining priority of invention there shall be considered not only the respective dates of conception and reduction to practice of the invention, but also the reasonable diligence of one who was first to conceive and last to reduce to practice, from a time prior to conception by the other.

The US Act also has a specific reference to plants (Sect. 161): 'Whoever invents or discovers and asexually reproduces any distinct and new variety of plant, including cultivated sports [a plant or animal deviating suddenly or strikingly from the normal type], mutants, hybrids, and newly found seedlings, other than a tuber propagated plant or a plant found in an uncultivated state, may obtain a patent therefor, subject to the conditions and requirements of this title. The provisions of this title [35 USCS Sects. 1 et seq.] relating to patents for inventions shall apply to patents for plants, except as otherwise provided.'

Further, Section 162 states, 'No plant patent shall be declared invalid for noncompliance with section 112 of this title [35 USCS Sect. 112] if the description is as complete as is reasonably possible. The claim in the specification shall be in formal terms to the plant shown and described.' And Section 163 states: 'In the case of a plant patent the grant shall be of the right to exclude others from asexually reproducing the plant or selling or using the plant so reproduced.'

Those are the main relevant sections of the Acts. Now for the historical path to patenting life. Anyone interested in a more detailed history than is provided in this chapter will find Mark Perry's article, 'From Pasteur to Monsanto: approaches to patenting life in Canada' (Perry, 2008), a useful starting place. Although focused on Canada, the scope is broader than Canada and it provides references to other jurisdictions.

Even though issues around patenting living things go back to at least Pasteur, a common starting point of histories of the legal evolution of patent on plants and animals is a 1969 German case concerning a red dove (Bundesgerichtshof, 27 March 1969 (German Federal Supreme Court) 1 I.I.C. 136 (*Rote Taube*)). The patent was denied because the 'specification' in the disclosure did not meet the requirements of the Act. What makes this case interesting is that in the judgement the Court expanded the interpretation of what can be patented to include biological entities. This expanded interpretation was explicitly cited in a 1976 German Federal Supreme Court judgement on the patentability of computer programs. In that judgement, the Court stated:

> According to the Wettschein (betting certificate) decision of the Federal Court, a technical invention is present, if an instruction is given to solve a technical problem by using specific technical means to achieve a technical result. In the Rote Taube (red dove) decision, this court generalised this definition so as to accommodate other forces of nature than those of physics and chemistry, e.g. those of biology. However in all cases the plan-conformant utilisation of controllable forces of nature has been named as an essential precondition for asserting the technical character of an invention. (Bundesgerichtshof, 22 June 1976) (German Federal Supreme Court), English version available at: http://eupat.ffii.org/papers/bgh-dispo76/index.en.html

This made possible the patenting of a biotechnological process and/or product as long as the plan (method) for the utilisation of the forces of nature (biological in this case) is a necessary element in the technical character of the invention. The court effectively ruled that the distinction between living and non-living was not a relevant consideration.

The first landmark decision in the USA was rendered by the US Supreme Court in the case of *Diamond* v. *Chakrabarty*, 447 U.S. 303 (1980). Sidney A. Diamond, Commissioner of Patents and Trademarks, was appealing to the Supreme Court the decision of the Court of Customs and Patent Appeals, which granted Arnanda Mohan Chakrabarty, who at the time was working for General Electric, a patent for a bacterium containing two energy plasmids which degraded oil. In its ruling the Court of Customs and Patent Appeals claimed, 'The fact that micro-organisms are alive is without legal significance for purposes of the patent law.' This echoes the earlier statement of the German Federal Supreme Court. The Supreme Court dismissed the appeal, deciding in favour of Chakrabarty. The Supreme Court held:

> A live, human-made micro-organism is patentable subject matter under § 101. Respondent's micro-organism constitutes a "manufacture" or "composition of matter" within that statute. Pp. 447 U.S. 308–318.
>
> (a) In choosing such expansive terms as "manufacture" and "composition of matter," modified by the comprehensive "any," Congress contemplated that the patent laws should be given wide scope, and the relevant legislative history also supports a broad construction. While laws of nature, physical phenomena, and abstract ideas are not patentable, respondent's claim is not to a hitherto unknown natural phenomenon, but to a nonnaturally occurring manufacture or composition of matter – a product of human ingenuity "having a distinctive name, character [and] use." *Hartranft v. Wiegmann*, 121 U.S. 609, 121 U.S. 615. *Funk Brothers Seed Co. v. Kalo Inoculant Co.*, 333 U.S. 127, distinguished. Pp. 447 U.S. 308–310.
>
> (b) The passage of the 1930 Plant Patent Act, which afforded patent protection to certain asexually reproduced plants, and the 1970 Plant Variety Protection Act, which authorized protection for certain sexually reproduced plants but excluded bacteria from its protection, does not evidence congressional understanding that the terms "manufacture" or "composition of matter" in § 101 do not include living things. Pp. 447 U.S. 310–314.
>
> Page 447 U. S. 304
>
> (c) Nor does the fact that genetic technology was unforeseen when Congress enacted § 101 require the conclusion that micro-organisms cannot qualify as

> patentable subject matter until Congress expressly authorizes such protection. The unambiguous language of § 101 fairly embraces respondent's invention. Arguments against patentability under § 101, based on potential hazards that may be generated by genetic research, should be addressed to the Congress and the Executive, not to the Judiciary. Pp. 447 U.S. 314–318.
>
> 596 F.2d 952, affirmed.

Given this decision and the reasoning of the German Federal Supreme Court in 1969 and the US Supreme Court in 1980, the stance taken by the Supreme Court of Canada nine years later (*Pioneer Hi-Bred Ltd.* v. *Canada (Commissioner of Patents)*, [1989] 1 S.C.R. 1623) seems out of step. The wrestling with the patentability of life by the Canadian courts captures well the central concerns expressed by many about patenting life, so it is worthy of detailed attention. The concern it does not address and to which I will return in Section 6.1 is the corporate control of life forms.

The Supreme Court of Canada judgement begins with the claim:

> The real issue in this appeal is the patentability of a form of life. This is in fact a claim for a new product developed in the field of biotechnology, an area of activity taking in all types of techniques having a common purpose, "the application of scientific and engineering principles to the processing of materials by biological agents to provide goods and services" (A. T. Bull, G. Holt and M. D. Lilly, *Biotechnology: International Trends and Perspectives* (1982), at p. 21).

It also set out clearly the nature of the biotechnology involved in the Pioneer Hi-Bred case:

> Genetic engineering can occur in two ways. The first involves crossing different species or varieties by hybridization, altering the frequency of genes over successive generations. The main consequence of this intervention is to oppose within the same cell allelic genes, that is, opposing characteristics which replace each other alternately in the hereditary process, as a consequence of the alternate action of their dominant genes. Naturally, the genes only offer a reasonable prospect that the traits will be acquired from one generation to the next. It should further be remembered that acquiring a certain characteristic does not automatically mean developing that characteristic: some effects in gene development and the influence of environment can cause genetic mutations. Besides it appears that "[V]arious studies indicate that mutations take place at random in time and space, having no relation to possible survival value" (N. M. Jessop, *Biosphere: A Study of*

Life (1970), at p. 294). There is thus human intervention in the reproductive cycle, but intervention which does not alter the actual rules of reproduction, which continues to obey the laws of nature.

This procedure differs from the second type of genetic engineering, which requires a change in the genetic material – an alteration of the genetic code affecting all the hereditary material – since in the latter case the intervention occurs inside the gene itself. The change made is thus a molecular one and the "new" gene is thus ultimately the result of a chemical reaction, which will in due course lead to a change in the trait controlled by the gene. While the first method implies an evolution based strictly on heredity and Mendelian principles, the second also employs a sharp and permanent alteration of hereditary traits by a change in the quality of the genes.

The genetic engineering performed by Hi-Bred is of the first type. Hi-Bred obtained this new soybean variety by hybridization, that is by crossing various soybean plants so as to obtain a unique variety combining the desirable traits of each one. This is why, as the Hi-Bred patent application explains, selective reproduction was necessary after crossing: making the new line grow, keeping only plants with the desired characteristics and repeating the operation for a sufficient number of generations to ensure that the soybean plants will finally contain only genes having the ideal traits. In this connection I would mention that the passages included in evidence in the record of the Court by Hi-Bred give a good idea of the various procedures used to obtain improved soybean varieties.

The Hi-Bred argument rests on the particular characteristic of the reproductive cycle of the soybean. The male and female gametes are contained in the flower and are protected from almost any intrusion at the time of reproduction. "Artificial" intervention is thus necessary to alter the cycle. The scope for "natural" crossing is therefore almost nil. Appellant argued that human intervention and the innovative nature of this new variety are conclusive and allow it to "qualify" for a patent under the *Patent Act*.

On the central matter, it declared:

The intervention made by Hi-Bred does not in any way appear to alter the soybean reproductive process, **which occurs in accordance with the laws of nature** [emphasis added]. Earlier decisions have never allowed such a method to be the basis for a patent. The courts have regarded creations following the laws of nature as being mere discoveries the existence of which man has simply uncovered without thereby being able to claim he has invented them.

> Hi-Bred is asking this Court to reverse a position long defended in the case law. To do this we would have, *inter alia*, to consider whether there is a conclusive difference as regards patentability between the first and second types of genetic engineering, or whether distinctions should be made based on the first type of engineering, in view of the nature of the intervention. The Court would then have to rule on the patentability of such an invention for the first time. The record contains no scientific testimony dealing with the distinction resulting from use of one engineering method rather than another or the possibility of making distinctions based on one or other method. (emphasis added)

Ultimately, the Court dismissed the appeal, stating:

> In view of the complexity presented by the question as to the cases in which the result of genetic engineering may be patented, the limited interest shown in this area by the parties in their submissions, and since I share the view of Pratte J. that Hi-Bred does not meet the requirements of s. 36(1) of the Act, I choose to dispose of this appeal solely on the latter point.
>
> . . .
>
> Having found that there was not sufficient disclosure of this soybean variety and that it therefore cannot be a patentable matter within the meaning of the *Patent Act*, it is neither necessary nor desirable for the reasons already given to consider in this appeal whether this new soybean variety can be regarded as an invention within the meaning of s. 2. I would accordingly dismiss the appeal. There shall be no adjudication as to costs.

This is somewhat convoluted reasoning, at least so it seems to me, so let's unravel some of its threads. First, the appeal was dismissed on the ground of a failure to adequately disclose, so the additional commentary of the Court does not really underpin the dismissal. Second, Pioneer Hi-Bred was not attempting to patent a product of molecular engineering but rather of hybridisation, which would normally fall under a legislated 'plant varieties protection' (in Canada, Plant Breeders' Rights Act (1990) and regulations which provide legal protection to plant breeders for new plant varieties for up to 18 years). The Court noted the difference in the engineering techniques (molecular genetic techniques and population genetic techniques) and was clear that whether, in the context of patentability, there was a conclusive difference had yet to be decided. It chose not to address that issue but was clear that the form of life arising from population genetic techniques was not patentable, because, the 'soybean reproductive process... **occurs in accordance with the laws of nature'** (emphasis added).

This takes us to the third thread, namely, what sort of appeal to the laws of nature is being made here? A chemist who developed a product by finding a new fractionation of oil is not in any way violating or transcending 'the laws of nature'. Nonetheless, she would be able to patent this process and product. Indeed, every invention in the realm of physics and chemistry occurs in accordance with the laws of nature. Otherwise, patentability would require the demonstration of a miracle. This suggests that the reproductive processes of forms of life are to be construed differently in patent law than chemical processes or the processes governing, say, the behaviour of subatomic particles. Since we are attempting to understand a claim made in 1989 – not one made in 1900 – it is appropriate that we focus on 1989 scientific knowledge. Given that molecular processes underlie the behaviour of chromosomes and other aspects of the reproductive process, it would seem that a distinction between the molecular and chromosomal level with respect to reproductive processes is untenable. Furthermore, molecular biological processes, including reproductive processes, are entirely describable biochemically, making a distinction between chemical processes and molecular processes untenable. So, if inventions based on chemical processes that occur in accordance with the laws of nature are patentable, it is not easy to discern the case for claiming that those based on reproductive processes are not.

Perhaps, and this is our fourth and final thread, the Court was making a very narrow claim, one applicable only to the specific invention and related processes involved in the Pioneer Hi-Bred case (and others similar to it). Perhaps the comment really has little to do with the soybean reproductive process, which occurs in accordance with the laws of nature. Pioneer Hi-Bred claimed that the new variety could not have arisen without human intervention because of features of the soybean's reproductive processes: 'the male and female gametes are contained in the flower and are protected from almost any intrusion at the time of reproduction'. The Court's reasoning might have been that all Pioneer Hi-Bred did was manipulate a contextual feature of reproduction. This would be analogous to simply lowering the temperature to below that in which a catalytic chemical reaction would have otherwise taken place in accordance with the laws of nature. The first thing to note is that if the Court focused on a special feature of this specific reproductive process, then its comments cannot be generalised to other biotechnical processes. But more significantly, the reasoning still appears specious.

The hurdle the Court imposed was that patentability required a change in the plant's reproductive processes such that the new process was not in accordance with the laws of nature. Since it would be absurd for the Court to be demanding a miracle – a breaking of the laws of nature – let's assume 'obeys laws of nature' is an ill-chosen phrase and what is intended is that a patentable invention requires some 'control' over the process of reproduction in the plant and not just over the conditions under which reproduction occurs. An example might be a chemical that when applied to the leaves of plants reliably causes gametes to combine in specific, non-normal ratios during reproduction. The laws of nature are intact but the actual 'natural' course of reproduction has been 'controlled' so that it occurs differently, obeying different 'laws of nature'. I admit that the Court's claim, 'There is thus human intervention in the reproductive cycle, but intervention which does not alter the actual rules of reproduction, which continues to obey the laws of nature', does seem at odds with this interpretation, but let's let that go.

Since a hurdle in this case can reasonably be expected to be a hurdle in all cases, what does this say about non-living inventions? Suppose someone finds a special process, which keeps all particulate matter off a freshly varnished surface. The importance of this to achieving an uncontaminated surface that is smoother, stronger and long-lasting may be great in the manufacture of furniture. The technique does not change in any way the varnish or the laws of nature governing its drying and hardening (its curing). It simply manipulates through a clever process the particulate matter in the surrounding air in just the way Pioneer-Hi-Bred did with the pollen particles. Consistent application of the Court's reasoning seems to entail that this invention is unpatentable, a finding which does not seem consistent with actual case decisions in Canadian patent law decisions.

This decision was handed down on 22 June 1989; on 1 October 1989, a new Patent Act came into effect. The changes in the patent act do not address the issues raised in this decision. The Commissioner of Patents issued a patent (CA 1313830) to Monsanto Technologies LLC (United States) on 23 February 1993 for 'GLYPHOSATE-RESISTANT PLANT CELLS' (Monsanto filed the patent application on 6 August 1986). Monsanto had already been issued a patent in the USA. The validity of the Canadian patent was challenged in a dispute between a Canadian farmer, Percy Schmeiser, and Monsanto. Roundup Ready canola was discovered growing in fields owned by Schmeiser. The seed had not been purchased from Monsanto; Monsanto sued Schmeiser for patent infringement. In the initial court case, the trial judge found the patent to be valid and ruled that Schmeiser had infringed the patent. He appealed the decision. On appeal, the Federal Court of Appeal affirmed the decision but made no finding on patent validity. Schmeiser then appealed to the Supreme Court of Canada.

The Supreme Court focused on three questions: whether the patent is valid, whether the patent was infringed, and whether the remedy was appropriate. In a 5 to 4 decision, it declared the patent valid, that an infringement had occurred but that the remedy was inappropriate:

> 94 Our task, however, is to interpret and apply the *Patent Act* as it stands, in accordance with settled principles. Under the present Act, an invention in the domain of agriculture is as deserving of protection as an invention in the domain of mechanical science. Where Parliament has not seen fit to distinguish between inventions concerning plants and other inventions, neither should the courts.
>
> 97 We conclude that the trial judge and Court of Appeal were correct in concluding that the appellants "used" Monsanto's patented gene and cell and hence infringed the *Patent Act*.

On the issue of restitution and remedy, the Court held:

> Their [Schmeiser and his contracted growers] profits were precisely what they would have been had they planted and harvested ordinary canola. They sold the Roundup Ready Canola they grew in 1998 for feed, and thus obtained no premium for the fact that it was Roundup Ready Canola. Nor did they gain any agricultural advantage from the herbicide resistant nature of the canola, since no finding was made that they sprayed with Roundup herbicide to reduce weeds. The appellants' profits arose solely from qualities of their crop that cannot be attributed to the invention.
>
> 105 On this evidence, the appellants earned no profit from the invention and Monsanto is entitled to nothing on their claim of account.
>
> IV. Conclusion
>
> 106 We would allow the appeal in part, setting aside the award for account of profit. In all other respects we would confirm the order of the trial judge. In

view of this mixed result, we would order that each party bear its own costs throughout.

Four justices dissented from this judgement. It is worth noting that the dissenting justices were not in complete agreement; Justice LeBel dissented in part from the overall dissenting argument. Interestingly, at least to me, one of the dissenting justices in this case was the Honourable Frank Iacobucci, a person of impeccable integrity, in my view. In the Pioneer Hi-Bred case, Dr Iacobucci was the solicitor for the respondent (i.e. for the Commissioner of Patents). At the time the Pioneer Hi-Bred case was argued before the Supreme Court, he was Deputy Minister of Justice in the Federal Government – a position he held from 1985 to 1988 – and, hence, was arguing the position of the government of the day. His reasoning and the principles to which he alludes in the Schmeiser case are much the same as those he adduced in the Pioneer Hi-Bred case.

The reasoning of the majority of justices is more important than the specific decisions because the reasoning parallels the international community's reasoning more closely than the Pioneer Hi-Bred reasoning. It is also one of the most detailed explications of this line of reasoning. In reaching its judgement, the Court acknowledged that Monsanto did not, and indeed could not, hold a patent on the plant that contained the gene for the Roundup trait. This is consistent with a previous Supreme Court decision in the 'Harvard mouse case' (*Commissioner of Patents* v. *President and Fellows of Harvard College*, 2002 SCC 76). This was a somewhat surprising decision. In a commentary on this decision, published in *Bio Business* (2003), Cynthia Tape and Conor McCourt wrote, 'On December 5th (2002), the Supreme Court of Canada made a landmark ruling that every biotechnology inventor and investor should know about: higher life forms – including multicellular differentiated organisms such as plants, seeds and animals – are not patentable subject matter in Canada.' They also noted that the decision 'sets Canada apart from much of the industrialised world, which permits higher life forms to be patented' (Tape and McCourt, 2003).

In this case, the Court held that even though Monsanto did not hold a patent on the plant that contained the gene for the Roundup trait, its patent rights were violated nonetheless because patented parts used in a whole entity are still protected. If a car manufacturer employed a new, patented fuel-injection system in its cars, it could not argue that it had not violated the fuel-injection patent because it was just one of many parts in a vehicle for which it, the car manufacturer, held the patent. This decision brought Canadian patent law closer to the emerging international consensus.

At this point, in the countries with the major biotechnological industries, the legal question of the patenting of life has been settled. However, the public debate continues. Two major issues that are in dispute are the morality of allowing living things to be patented and the impact of allowing private corporations to hold such patents. These are entirely separable issues since one could accept that life can, from a moral perspective, be patented without accepting that such patents can and should be held privately. The debate about private corporations holding patents on living things is explored in Section 6.1. The force of the claim that patenting life is ethically improper rests mostly on the ability to make a distinction between living things and non-living things. This distinction was explored in Section 4.2.

5 The controversy

Purported benefits

Since 2005, when the first commercial plantings of GM crops occurred, farmers have been the principal immediate beneficiaries: higher yields, lower input costs and so on. There have been benefits to consumers and to the environment but these are less visible. Consumers have benefited from a secure supply of food at stable or falling prices – even though, for example, the cost of oil-derived products (e.g. fuel – used in tractors and transport trucks – and artificial fertilisers) has increased significantly. The environment has benefited from reduced pesticide and herbicide spraying, reduced use of fuel in tractors (fewer herbicide sprayings required and no pesticide spraying), lower groundwater contamination, and zero tillage (reducing wind and water erosion of soils). There are, of course, claims of harms associated with GM crops as well as challenges to the claimed benefits; these are examined in the next chapter. The next generation of GM crops promises to have more tangible benefits for consumers (e.g. higher expression of specific vitamin enrichment and **long-chain** Ω-3 fatty acids – an important cardiac health benefit), for the environment (e.g. draught tolerance – hence, less irrigation water use – and lower fertiliser requirements – nitrogen fertiliser, for example, which is a significant source of greenhouse gases) and for farmers in low- and middle-income countries. In this chapter, I explore three purported benefits of GM crops.

5.1 Environment benefits

Farming (animal and crop) is exceptionally hard on the environment. According to the World Wildlife Fund (WWF) website, current agriculture uses about 50 per cent of habitable land, consumes 69 per cent of all water used by humans (municipal uses, including household use, comprise 8 per cent, and the total industrial use is 23 per cent), and is a significant contributor to greenhouse gases. Citing the Food and Agriculture Organisation of the United Nations, the WWF reports that livestock agriculture alone is responsible for 18 per cent of all greenhouse gas production, and rice production is the largest single producer of methane. Agriculture is also the largest user of chemicals – more than any industrial use. The WWF (www.worldwildlife.org/cci/agriculture.cfm) states:

> Agriculture is the leading source of pollution in many countries; in the U.S. alone, 428,200 metric tons of pesticides are introduced to the environment every year. Many of these pesticides are suspected of disrupting the hormone messaging systems of people and wildlife.

The WWF is quick to point out that agriculture provides many benefits and opportunities for environmental protection. Nonetheless, the picture it paints is depressing; current agricultural practices are environmentally destructive and unsustainable. In light of this, any claim that a technology can reduce this impact without negatively affecting the supply of food, its price or the health of consumers, is worthy of serious examination. Kalaitzandonakes (2003) provides some excellent analyses of economic and environmental impacts.

Genetic modification of crop plants is currently dominated by two traits: glyphosate (Roundup) resistance and *Bt* δ-endotoxin expression. Although these two predominate, there are other important GM traits, as, for example, in tomatoes in which an enzyme that causes tomatoes to rot and degrade is blocked, in papaya engineered to have virus resistance, and in tobacco engineered to produce low or no nicotine. Glyphosate is a broad-spectrum (non-specific) herbicide; almost all plants will be negatively affected by contact with it, most being killed. Through genetic modification, certain crops are made resistant to glyphosate. Hence, a farmer can plant these GM crops, spray the field with glyphosate, and kill the weeds without harming the crop plants. Until recently, Monsanto held the patent on glyphosate under the trademark Roundup, and the crops were designated 'Roundup ready'. Glyphosate, with respect to environmental and health impacts, compares favourably with alternatives such as atrazine (widely used on non-GM cornfields) and 2,4-D amine (commonly used in non-GM soybean fields. The US Environmental Protection Agency (EPA) in a Consumer Factsheet (www.epa.gov/ogwdw/pdfs/factsheets/soc/glyphosa.pdf – all following quotations from this source) claims:

Glyphosate is strongly adsorbed to soil, with little potential for leaching to ground water. Microbes in the soil readily and completely degrade it even under low temperature conditions. It tends to adhere to sediments when released to water. Glyphosate does not tend to accumulate in aquatic life.

Microbial activity and other chemicals may break down atrazine in soil and water, particularly in alkaline conditions. Sunlight and evaporation do not reduce its presence. It may bind to some soils, but generally tends to leach to ground water. Atrazine is not likely to be taken up in the tissues of plants or animals.

2,4-D is readily degraded by microbes in soil and water. Leaching to ground water may occur in coarse-grained sandy soils with low organic content or with very basic soils. In general little runoff occurs with 2,4-D or its amine salts. There is no evidence that bioconcentration of 2,4-D occurs through the food chain. This has been known from large-scale monitoring studies of soils, foods, feedstuffs, wildlife, human beings, and from other environmental cycling studies.

So it is reasonable to consider glyphosate to have a better overall profile than its alternatives with respect to environmental and health issues. Moreover, fewer applications of glyphosate are required on GM crop fields. This lessens the environmental herbicide load and decreases fuel consumption. Additionally, as already mentioned, a field can be planted with 'Roundup ready' crops without tilling the soil (zero tillage). This avoids exposing the soil to wind and water erosion (loss of topsoil from erosion due to farming practices is a major environmental concern) and, again, reduces fuel consumption. These environmental gains are realised while at the same time increasing per acre yields. The GM 'Roundup ready' crops used in North America are soybean, canola, corn (maize), tobacco and cotton.

One obvious environmental benefit of pest-resistant crops is that **no** pesticides are applied, eliminating numerous environmental impacts including groundwater contamination and hydrocarbon emissions from tractor use. δ-Endotoxins are naturally expressed by the bacterium *Bacillus thuringiensis*, which is ubiquitous in soils (see pp. 31–32). *Bt*, usually as a spray, has been used for more than 50 years as an insecticide – including controlling mosquitoes – and has an impeccable safety record; that record has led to its being widely used in organic farming and to the USA's EPA not imposing any requirements on it with respect to food residue tolerances, groundwater, animal or human toxicity labelling, or special reviews. Researchers at the University of California, San Diego, drawing on Glare and O'Callaghan (2000), claim:

> *Bt* products are found to be safe for use in the environment and with mammals. The EPA (environmental protection agency) has not found any human health hazards related to using *Bt*. In fact the EPA has found *Bt* safe enough that it has exempted *Bt* from food residue tolerances, groundwater restrictions, endangered species labeling and special review requirements. *Bt* is often used near lakes, rivers and dwellings, and has no known effect on wildlife such as mammals, birds, and fish.
>
> Humans exposed orally to 1000 mg/day for 3–5 days of *Bt* have showed no ill effects. Many tests have been conducted on test animals using different types of exposures. The results of the tests showed that the use of *Bt* causes few if any negative effects. *Bt* does not persist in the digestive systems of mammals.
>
> *Bt* is found to be an eye irritant on test rabbits. There is very slight irritation from inhalation in test animals which may be caused by the physical rather than the biological properties of the *Bt* formulation tested.
>
> *Bt* has not been shown to have any chronic toxicity or any carcinogenic effects. There are also no indications that *Bt* causes reproductive effects or birth defects in mammals.
>
> *Bt* breaks down readily in the environment. Because of this *Bt* poses no threat to groundwater. *Bt* also breaks down under the ultraviolet (UV) light of the sun.
>
> Even with such widespread use of *Bt*-based products in the past 50 years, only two incidents of allergic reaction have been reported to the EPA. In the first incident, it was concluded that the exposed individual was suffering from a previously diagnosed disease. The second involved a person that had a history of life-threatening food allergies. Upon investigation, it was found that the formulation of *Bt* also contained carbohydrates and preservatives which have been implicated in food allergy. (www.bt.ucsd.edu/bt_safety.html)

Given this profile of almost no environmental and health impact, *Bt* is a dream insecticide, which is why, in spray form, it is widely used by **organic** farmers.

The next generation of GM crops, many very close to being marketed, includes draught tolerance, reduced nitrogen and enhanced nutrients. Drought tolerance will result in reduced water use. Figure 5.1 provides data on yield improvement for GM corn under drought stress. This trait will increase in

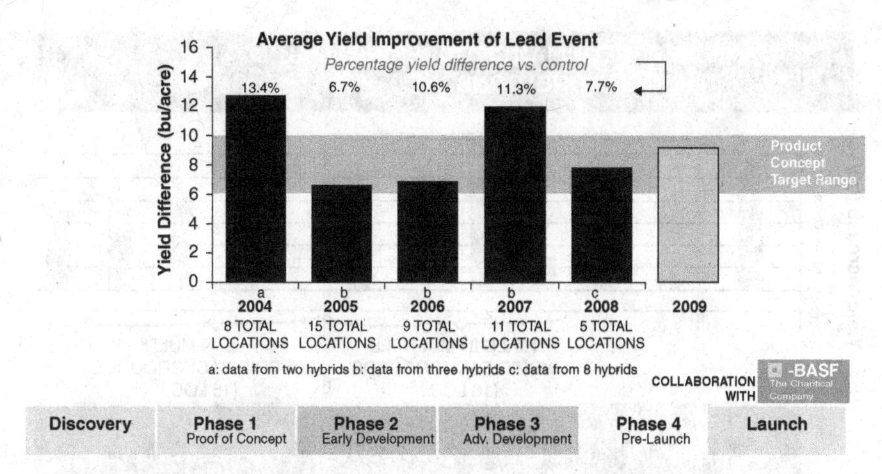

Figure 5.1 Performance of cspB corn in USA across years (fifth season of yield improvements under drought stress) (courtesy of Monsanto).

importance as microclimate changes occur (see Fay and Bierbaum, 2010, who co-directed *The World Development Report 2010: Development and Climate Change*).

Reduced nitrogen requirements will have numerous environmental benefits from less use of fossil fuel in the production of synthetic nitrogen fertiliser to lower emissions into the atmosphere from eventual gassing off from lakes and rivers. Figure 5.2 provides data on yields for GM nitrogen-reduction plants compared to controls with the same nitrogen inputs, and Figure 5.3 shows the reduction-in-nitrogen to yield ratios.

Enhanced nutrient profile is, obviously, a benefit to human health but might also result in a reduction of agricultural land use as a result of fewer sources of essential nutrients being required.

The current and immanent environmental benefits are easily understood (see http://oecdinsights.org/2010/04/19/ge-crops-good-for-farmers-good-for-the-planet/ and http://news.stanford.edu/news/2010/june/agriculture-global-warming-061410.html). The question that critics focus on is, 'at what cost?' Are there harms to the environment and health that have to be balanced against these benefits? Not surprisingly, there are and they are discussed in the next chapter.

5.2 Yield and food security benefits

Rich countries have not had to face issues of food security for the last 50 years. Food has been abundant, safer than at any previous time, and inexpensive

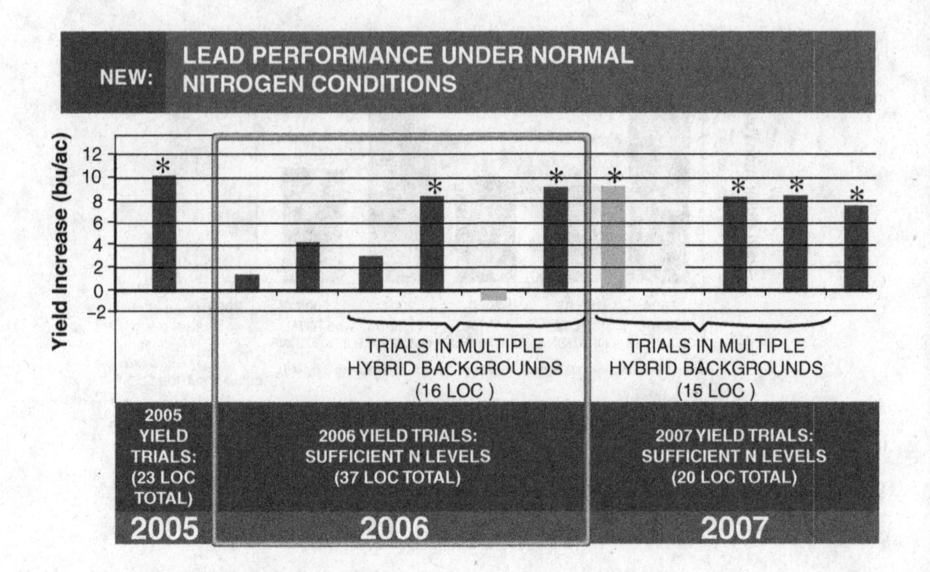

NEW: LEAD PERFORMANCE UNDER NORMAL NITROGEN CONDITIONS
Yield Increase (bu/ac)
12
10
8
6
4
2
0
-2
TRIALS IN MULTIPLE HYBRID BACKGROUNDS (16 LOC)
TRIALS IN MULTIPLE HYBRID BACKGROUNDS (15 LOC)
2005 YIELD TRIALS: (23 LOC TOTAL)
2006 YIELD TRIALS: SUFFICIENT N LEVELS (37 LOC TOTAL)
2007 YIELD TRIALS: SUFFICIENT N LEVELS (20 LOC TOTAL)
2005
2006
2007
* Statistically significant @ps0.10
Bar color correlates with the specific hybrid background tested. Same bar color in different tests and different years indicates same hybrid was used.
All trials conducted under sufficient nitrogen application levels.

Figure 5.2 Nitrogen trials (courtesy of Monsanto).

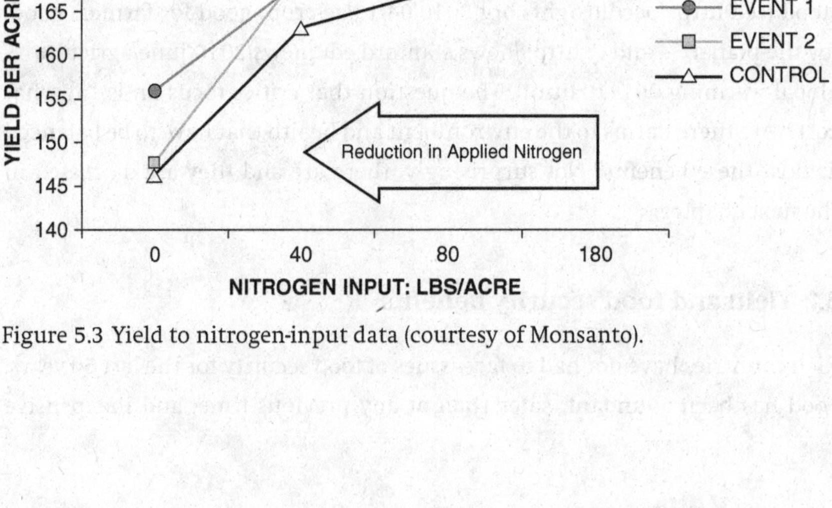

2006 FIELD RESULTS CONFIRM CONTINUED PERFORMANCE OF LEADS IDENTIFIED IN 2005
LEAD NITROGEN UTILIZATION GENE (ACROSS 3 LOCATIONS: ILLINOIS AND IOWA)
YIELD PER ACRE
175
170
165
160
155
150
145
140
EVENT 1
EVENT 2
CONTROL
Reduction in Applied Nitrogen
0
40
80
180
NITROGEN INPUT: LBS/ACRE

Figure 5.3 Yield to nitrogen-input data (courtesy of Monsanto).

relative to income; although some individuals in rich countries do not have adequate access to food, the main factor is poverty, not food availability, safety or reasonableness of cost. Things have not always been this way; famines were common in all regions of the globe prior to the twentieth century. Today, famine largely befalls low-income countries. Crop failures do occur in rich and middle-income countries, but global production and trade mitigate their impact, as does the diversity of agricultural crops and animals. During, and for a brief period after, the wars of 1914–1918 and 1939–1945, some rich countries faced issues of food security and many foods were rationed. The depression of the 1930s also caused food security problems. The last 50 years of abundance have dulled memories and created a potentially flawed sense of optimism. Food and water top the list of necessities of life; it is unwise to become sanguine about their continued availability.

The current abundance of food is in significant part the result of advances in biology and specifically biotechnological applications. Yields have been steadily increasing over the last century as a result of significant advances in our knowledge of plant and animal nutritional requirements, determinants of optimum germination, insemination and gestation, plant and animal diseases and how to avoid or treat them, and plant and animal genetics. It is also, in part, a result of global food markets; a famine in one area of the world is mitigated by abundance elsewhere.

Figure 1.1 (see p. 16) indicates the significant yield increases for maize since the early 1930s. The slope of the line shows the rate at which yields have increased. The steeper the slope, the faster yields are increasing year-over-year. The slope of the line graphing the period of open pollination is virtually flat, meaning little if any increases in yields were occurring. Beginning in 1931, with the introduction of biotechnology-based double crosses (see Section 1.3 above), the slope increases dramatically (with yields increasing 63.1 kg per hectare per year). In 1959, with introduction of single crosses (the next biotechnical advance), the slope increased again (with yield increases of 113.2 kg per hectare per year). In 1995, GM maize was introduced and the slope increased yet again (with yield increases of 207.2 kg per hectare per year). These increases resulted from genetic manipulation (hybridisation in the first two cases and molecular manipulation in the third) and changes in 'cultural practices'. Cultural practices include fertiliser use and the timing of its application (whether manure or synthetic fertiliser), pesticide and herbicide use, crop rotation, zero tillage and irrigation. Most of those practices were in place by 1995, so the additional yearly increases after 1995 are almost entirely the result of genetic biotechnology.

As the graph indicates, during the open-pollination period the yield was 1,600 kg per hectare per year; today it is 9,400 kg per hectare per year. As already noted 22 per cent fewer hectares are planted in maize today than in 1931. A similar increase-in-yield pattern is found for soybean, canola and wheat. Current total production of these crops frequently results in overabundance – not rampant overabundance such that prices plummet but enough to keep prices moderate and provide some margin of safety. Hence, some decrease in production is tolerable, which, in rich countries, makes eschewing GM crops tolerable; without GM maize, for example, yearly production per hectare would still be about 8,800 kg. Reverting, however, to 1931, the open-pollination period, levels of production of soybeans, wheat and maize (1,600 kg per hectare for maize) would be catastrophic. The prices of these commodities would rise dramatically as would food products from bread to beef (soybean is a major ingredient in cattle feed) to cooking oils. Only about 20 per cent of the current supply of these commodities and all the associated food products would be available, so many would have to do without or everyone would have access to only 20 per cent of today's supply. Without genetic manipulation, abundant and affordable food would evaporate. A line of thought that surfaces occasionally is that these dramatic yield increases have been harmful because they have made possible a population explosion, which is overburdening the planet in numerous ways. Quite the opposite is, in fact, the case. In rich countries, a rise in affluence has been accompanied by a declining birth rate, resulting in a stable population size since 1950. The population growth has for the last 60 years occurred (and will for the foreseeable future occur) in poor countries, where food is anything but plentiful and starvation is common.

An increase in yields is a major factor in achieving an abundant, affordable and secure food supply. The emergence of international agricultural markets is another. People in rich countries are not dependent on local circumstances for a reliable supply of food. A drought in an area has only a modest impact on availability and price, because agricultural commodities move rapidly over long distances to eliminate the shortage in the drought-ridden region. International markets also increase competition, thereby, moderating prices. The existence of these markets is essential to food security. That, however, does not entail that preferential support of local agricultural products is undesirable. The 'eat locally' movement (the 100-mile diet and locavore movement, for example) highlights the benefits (and harm reductions) of purchasing locally produced foods. Here, however, as in many other cases explored in this book, the prudent and ethically defensible position avoids the either-or trap. If everyone in rich countries ate only food produced within 100 miles of their homes, international markets would collapse. An exceptionally wet and cold growing season would result in substantially lower yields and in some cases crop failures. This would reduce the available food for people and agricultural animals, and prices would soar. There would be no international market solution. People in poor countries know this scenario well. So, that end of the spectrum is neither prudent nor ethically defensible; it is also highly unlikely to occur.

The other end of the spectrum, where all food in a region is obtained from international markets, is also not prudent or defensible. The market structure internal to the regions would collapse. Farmers, of course, would still sell in the international market place but local activity would cease. This undermines any self-sufficiency for the region and exposes it completely to external market forces. It is highly unlikely that this extreme will occur. Hence, as is almost always the case, the prudent and ethically defensible position is to strike the right balance.

Biotechnology and international agricultural markets (including processing industries) are essential to abundance and food security. However, both pose significant regulatory challenges, which we all too often address poorly. With respect to international markets, there is a tendency to overregulate or underregulate them, and, worse, to vacillate between these states; as governments change, so do political ideologies regarding government's regulatory role.

With respect to biotechnology, addressing the regulatory challenges is made more complicated by the rapid pace at which biological knowledge and its application are growing. Moreover, while some civil servants have a solid understanding of the science and technology, most politicians and the majority of citizens lack the requisite understanding of the science and technology. The media also seem less and less equipped to engage citizens in informed debate. Most media depend on advertising revenue, and attracting that revenue depends on audience size. Many things affect audience size but very low on the list is sustained discourse on science and technology; such discourse appeals to a boutique market, which is usually served by a media that is less reliant on audience size, such as public broadcasting, or attracts advertising revenue from those more interested in reaching a market sector. As a result, fewer and fewer of those who report on science and technology have the background and skills to dissect a research paper themselves and ferret out the important questions about, for example, experimental design, data interpretation and knowledge gaps that remain. And, all too often, those that do have the skills, no longer have the time or encouragement to employ them; they, also, are given inadequate media time or space to provide more than a cursory account.

Ultimately, politicians, not civil servants or research scientists, make major policy decisions; civil servants, of course, are inextricably involved in interpreting and applying them. Politicians make decisions with an eye on the ballot box; that is, with an eye on the attitudes and views of citizens. So, those least equipped to make such decisions and operating in a less than rational environment, stumble along. This is not unique to agricultural biotechnology; it pervades all science and technology. It is especially prominent in health research and its applications (stem cell research being a current high-profile issue); in medicine, however, genetic modification of bacteria, yeast and animals seems to fly under the public and political radar in a way agricultural genetic modification does not. The default political stance with respect to new applications of science is, 'no, except in this or that exceptional case'. This is seldom a sustainable stance. Putting technological genies back into their bottles has never been successful. Eventually (sometimes a decade or more later), the more productive stance is taken, 'yes, but not in these cases'. This is usually more a result of resignation to what has become entrenched than embracing and sensibly regulating a new technology. Frequently, in hindsight, it is hard to capture what all the fuss was about. This is regrettable because the benefits of informed public debate are lost, and for the most part regulations merely codify what has become accepted practice rather than shaping that practice in the early stages of its development.

5.3 Health benefits

The public visibility of health benefits from GM crop technology over the last 15 years has been poor. Nonetheless, some direct consumer health benefits have been realised such as some reduction of toxins in food. For example, the mould *Fusarium*, which infects corn, along with many other plants, synthesises trichothecenes (type B), T-2 toxin, zearalenone (F-2 toxin), vomitoxin, deoxynivalenol and fumonisin. The European corn borer compromises the cob and kernel integrity, increasing the entry points for the mould spores; *Bt* corn reduces the activity of the corn borer, hence reducing the compromised integrity of the cob and kernels. Munkvold and Hellmich (2000) published their research on this and claimed:

> Because the fungi that produce mycotoxins in maize are frequently associated with insect damage to the plants, insect control has the potential to reduce mycotoxin concentrations in grain. Here we summarize six years of research that indicate that Bt transformation of maize hybrids enhances the safety of grain for livestock and human food products by reducing the plants' vulnerability to mycotoxin-producing *Fusarium* fungi. Lower mycotoxin concentrations represent a clear benefit to consumers of Bt grain, whether the intended use is for livestock or human foods. (www.plantmanagementnetwork.org/pub/php/review/maize/)

Public visibility of health benefits is, however, about to improve dramatically. The major agricultural biotechnology companies are well into the research and development cycle of crops that enhance nutrition, crops, for example, whose oils contain Ω-3 fatty acids (found mostly in fish oils at present), which promote cardiac health. The list of nutritional enhancements is growing rapidly and currently includes vitamin-enriched foods (along the lines of golden rice) and plants that produce important therapeutic agents.

I predict that public attitudes to GM crops will become more positive with the introduction of such crops. As I point out in several places, health and longevity matter to people in rich countries. Preventing and treating disease and stalling death are pressing concerns to virtually everyone. In an environment of abundant food, genetic modifications that simply add to the abundance fail to resonate with consumers, so any whiff of harm is sufficient to raise questions about the wisdom of engaging in the genetic modification. Health is different; people immediately understand the obvious downsides of not exploiting genetic modification to prevent and treat diseases. Hence, the potential harms have to be significant in order to forgo the benefits to health. Plant (and animal – but more slowly) agriculture is about to enter the health benefits arena.

In terms of biofactories for the production of therapeutic agents, bacteria and yeast have been used the most. There are clear advantages to using them; both function in laboratory conditions, both multiply rapidly, and both are reasonably easy to genetically modify. Although agricultural plants are reasonably easy to genetically modify – of course, a lot of knowledge and skill is involved, nonetheless – they are grown in fields and rarely have more than two generations a year. Plants, however, have characteristics that make them a better choice for certain health interventions. First, although they require large tracts of land and lots of attention, they are already grown in abundance. So, the infrastructure for growing, harvesting, processing and distributing already exists and will continue to exist even without health-enhancing genetic modifications. The major cost is in the research and development of a GM plant. After that, its agricultural cycle simply piggybacks on the existing cycle, with few additional costs over non-GM agricultural plants.

Second, most agricultural plants or elements derived from them, such as oils, are consumed by people. This means that a health-promoting agent, such as vitamin D or Ω-3 fatty acids, can be delivered to everyone as part of a normal diet. This would be one more step in a process that began decades ago. The benefit of fluoride (in water, salt or toothpaste) in reducing dental cavities is now incontestable. (As with everything, there is no shortage of critics on this issue – just Google 'fluoridation' – but there is no controversy in the medical or dental communities.) Delivering it through the water supply (introduced in the 1950s in the USA) was a great public health benefit. The same is true of iodine added to salt, vitamins A and D added to milk, vitamin-enriched flour, and the list goes on. Engineering health-promoting agents into agricultural plants follows this public-health approach. One obvious advantage to this approach is that it cuts across the socio-economic spectrum; access is universal and not dependent on ability to pay, knowledge or regional disparities in availability. In addition, the need to remember to take a capsule is eliminated, and no actions, such as acquiring a capsule, are required, actions which frequently decrease patient compliance.

This approach to delivering health-promoting agents does raise ethical issues; individual choice can be diminished, and informed consent can be trammelled. David Castle (2006) provides an excellent exposition of the opportunities and ethical challenges of nutritional genomics. It is important to distinguish different grades of diminished choice. The greatest diminution of individual choice occurs when the addition of a purported health-enhancing agent is mandated by law, as in the case of water fluoridation in the USA.[1] All

[1] Adding vitamin A and D to milk might be a stronger case than fluoridation because there are alternative water sources readily available to consumers, but alternatives to enriched

publicly supplied water must be fluoridated. Informed consent and choice are not entirely overridden, even in this case, since individuals can still purchase non-fluoridated water or can treat the public supply to remove the fluoride. Both involve taking steps to avoid fluorine and impose additional costs. Given this, it would be difficult to mount an argument that individual choice had been eliminated; it might not even be possible to claim it has been compromised. Informed consent is more complicated.

Many consumers might not be aware that fluoride has been added, or if they do, they might not know its properties; so, making an *informed* decision is not possible. It would be facile to dismiss this by indicating that such consumers have only themselves to blame since the information is readily available. It is absurd to expect individuals to know everything or even to know when and what information to seek out. Even well-informed, well-educated and well-connected individuals know only a small fraction of the available information. The fluoridation case, however, is not an entirely clean example of diminished informed consent. Much of the water consumed around the world is naturally fluoridated – in many cases at levels higher than that added to North American water supplies. Hence, fluoride in water is a natural phenomenon. This could mean that a significant number of people are drinking 'contaminated' water or that natural North American water is deficient – with respect to human needs – and, hence, needs to be supplemented, in much the same way that vitamin C-deficient diets on eighteenth-century sailing ships led to scurvy, which was prevented by adding vitamin C to the diet. A justification of this 'correcting a deficiency' position could be based on evolution. Humans emerged in eastern Africa (current views focus on the Serengeti Plains or Ethiopia) where water is naturally fluoridated (Kilham and Hecky, 1973), so that would be the nutrient profile to which we adapted. Indeed, levels in water sources in east Africa are higher than those in North America. Given the high levels of fluorosis in east Africa, this increase may be more recent or fluorosis may be due to dietary changes (Kjellevold Malde *et al.*, 1997).

This is not a book on fluoridation but the lessons about informed consent and choice are relevant. Choice simply requires options; if pursuing an alternative is unreasonably demanding or expensive, then the alternative, in effect, does not exist. Informed consent requires information and an action – consenting. The need to acquire informed consent is appropriately raised

milk are limited, requiring specialised sourcing of milk. The fluoride case, however, has more political life (controversy) and allows the important issues to be raised.

when an intervention is contemplated that may have an impact (positively or negatively) on an individual's well-being. Different cases will yield different answers to the question of whether informed consent is required.

Returning now to agricultural plants that have been modified to contain health-promoting agents, with respect to consumer choice and informed consent, how close to water fluoridation or vitamin-enriched milk modification is agricultural plant modification? Not very close is the answer. The non-GM portion of the food industry is likely to remain robust. That sector is not going to be legislated out of existence, so if it does decline, it will be the result of weak consumer demand; that is, fewer and fewer consumers choosing to buy non-GM products. It would be a perverse sense of choice that required the production of something that few consumers were choosing, just to preserve an abstract notion of a potential choice. Businesses, unless constrained by legislation or regulation, follow consumer choices; even with a small boutique market of non-GM consumers, that niche will be filled. Of course, those consumers may pay a premium for their choice. Unlike the fluoridation case, information is less of an issue in the GM-enriched plants case. Water does not come out of a tap labelled 'fluoridated'; food products are required to reveal ingredients on the label and, increasingly, a wealth of nutritional information as well.

As a final point, it is worth noting that prohibiting GM plants with health-promoting value is equally choice diminishing and compromising of informed consent. I might want the value of the agent and, because of aggressive lobbying and fear mongering, be denied that option. I might be much better informed about the science, health benefits and potential harms than those who oppose such agriculture and, based on that information, I might choose the GM product, but that option has been taken away.

The goal should be to expand choice not restrict it. This is a case where allowing market dynamics to reveal consumer choices is beneficial. Clearly, markets need to be regulated, so that harmful products with no offsetting benefits are not introduced, so that harm mitigation is maximal, so that claims have an evidentiary basis and are not misleadingly expressed, so that price collusion does not occur, and so on. But, with those regulatory desiderata met, the market will reflect, and respond to, consumer choice. A marketplace of competing ideas will influence that choice. Needless to say, that also requires regulatory discipline; outrageous claims that lack clarity and evidence need to be exposed. Fortunately, open public discourse is usually self-correcting.

Ironically, one of the impediments to open, self-correcting public discourse is government secrecy. To repeat a point made earlier, the Right and the Left of the political spectrum both seem to have a penchant for secrecy and control of information, to the detriment of vigorous political and social public debate, and, ultimately, choice, informed consent and democracy.

6 The controversy

Purported harms

The potential benefits of GM agriculture are, for the most part, not at issue in the debate, although whether it is appropriate to call some of them benefits, as I and many others claim, is contested by some critics. The central focus of critics is on harms. Three broad factions can be discerned. One faction stridently opposes GM, claiming real and serious harms exist that outweigh any claimed benefits. As one would expect, there is a spectrum of views within this faction, ranging from characterising the harms as catastrophic to simply unacceptable even given the benefits. A second faction cautiously accepts that the benefits are significant enough to outweigh the harms. The positions of those in this faction encompass one or more of: (1) a reluctant willingness to endure the harms (or risk of harms) to secure the benefits, (2) an unconcerned acceptance of the harms (or risk of harms), (3) a belief that the harms are serious but still outweighed by the benefits, and (4) a belief that the harms are not significant. Those in the third faction consider a significant fraction of the claims about harms to be exaggerated or outright false – a large subset of this faction think the claims are no more than disinformation and propaganda.

Some of the perceived harms are broader than GM agriculture but amplified by it; some others are specific to GM agriculture. The broad harms involve the perceived negative impact of economic globalisation, the power of multinational corporations and agri-business conglomerates, and the overly rapid deployment of innovations in science and technology. Specific harms include loss of heritage plants, effects of GM crops on non-target species, development of resistance in target pests to the toxin a plant has been engineered to express, horizontal gene transfer (HGT), introduction of new allergens and carcinogens, and changes in nutrient bioavailability. Generally, the broad harms flow from economic structures and forces, and from the relentless, and bumpy, social transformation wrought by science and technology. These are dealt with in Section 6.1. Mostly, the specific harms focus on environmental harms (dealt with in Section 6.2) and health harms (dealt with in Section 6.3).

6.1 Economic and corporate harms

This bundle of harms is frequently cast in terms of individual alienation and impotence in the face of economic structures and forces that control our lives but which are beyond our control – economic structures designed to benefit the few at the expense of the many. That alienation and impotence extend to the relentless social transformations wrought by science and technology, transformations that also control and transform our lives, and which are also beyond our control, and, in many cases, comprehension. Abuses of power, naked greed and deliberate deceptions appear to many people to be standard fare in the economic and business realm, and the constant revelation of them in the media has heightened the sense of consumer victimisation and diminished confidence in the effectiveness of regulations and regulatory agencies. As a result, public receptivity to claims of corporate control, deception and malevolence is high; spurious and real charges are, regrettably, accorded equal credibility and equal media attention. Sometimes it even seems unnecessary to provide evidence about this or that company or its actions, since it is obvious that, by their nature, companies are insensitive, deceptive and controlling; it is taken to be woven into the very fabric of economic systems.

This set of public attitudes is regrettable. It allows disingenuous critics a free ride to domination of the debate, and it undermines rational, productive discourse, as well as socially, politically and ethically defensible decisions, policies and regulations. Exploiting a public distrust of corporations, governments and other institutions is not an acceptable substitute for providing evidence and marshalling arguments to support a set of claims in a specific case. In that spirit, let's work through the purported harms that arise from the existence and behaviour of large corporations in an open, *but regulated*, market structure.

One concern is the concentration in a few large companies of control over seed production and distribution. A corollary of this concern is the impact on small-hold (small-scale) farmers – especially in poor, rural areas of low- and middle-income nations.

There is cause for concern with monopolies and large corporate control of a resource. It is, however, a mistake and a distraction to focus on one sector in wrestling with this issue. The reality is that Monsanto and other seed companies are relatively small multinationals, and the control they can exert over socio-economic factors pales by comparison with companies like Bayer, Coca-Cola, Toyota and Unilever. Consider, for example, Unilever, which has an extensive food-manufacturing and distribution arm. It is vastly larger than Monsanto or Pioneer or Syngenta. Information from Monsanto's annual statements and Unilever's website (www.unilever.com/aboutus/introductiontounilever/unileverataglance/index.aspx) conveys this vast difference in size. In 2009, Unilever employed 163,000 people in 170 countries and had a worldwide turnover of €39.8 billion (US$53.9 billion). The worldwide gross revenue (before deducting operating costs) for Monsanto in 2010 was US$10.2 billion. Hence, Monsanto is only 20 per cent the size of Unilever. Unilever claims, 'On any given day, two billion people use our products.'

The point, however, is not that Monsanto does not matter just because it is a relatively small multinational, but that the issue of multinational socio-economic power and control is much larger than GM seed companies, and any meaningful attempt to tackle the potential negative socio-economic consequences of the actions of multinational companies will require attention to issues that are unrelated, for the most part, to the specific products involved. I say 'potential' because multinational companies have a vested business interest in not being *seen* to produce negative socio-economic consequences. Hence, most multinational companies equal or exceed the level of corporate responsibility found in national and community-based businesses. Of course, sometimes there are concerns about a company's product or actions. Companies, multinational or much smaller in scope, do act in ways that result in harms; sometimes they are negligent and culpable, such as the BP oil debacle in 2010 in the Gulf of Mexico. It should be obvious, however, that where there are issues with the products and actions, those issues need to be the focus; the scale of the particular company should only be included where the issues themselves justify it.

Large multinational corporations are with us for the long haul – indeed are unavoidable, probably essential, in a sustainable global economy. The need for national and international legal frameworks, regulations, and enforcement mechanisms is beyond dispute. There is no reason, however, to single out GM seed companies, and GM crop and food companies, as needing special attention. Companies, regardless of size, like individuals, can and do, deliberately or inadvertently, misbehave and inflict harm on others. Consequently, oversight, in protection of public interest, is essential. Failures of oversight usually rest in part with the company and in part with the relevant legislative and regulatory bodies, as was the case with GM potato produced by Cambridge Agricultural Genetics, as discussed above.

Turning now to the second socio-economic concern, namely, creating a dependency on seed companies for annual seed inputs. This concern is specifically relevant to seed companies but is considerably less significant than often claimed. In the rich countries, almost all farmers who buy **non-GM** seeds from a company are doing so to obtain a beneficial trait. Since the majority of those traits are quantitative traits and result from hybridisation, seeds that are collected from one season and planted the next will result in approximately 50 per cent of the crop not having the trait. As a result, almost all farmers in rich countries buy hybrid seeds every year from a company that guarantees the presence of the trait in virtually every seed. Why would a farmer endure a situation where in every season 50 per cent of her crop lacks the trait from which she hopes to benefit, especially when, on that 50 per cent, she will incur all the costs of planting, fertilising and dealing with weeds and pests? To add to the complications and costs, the 50 per cent will not be neatly confined to a 1/2 section of the field; it will be entirely mixed with the plants manifesting the trait. I have a small vegetable garden at my home. I always buy new seeds every year. And I do so even for the non-hybrid heritage crops I grow, because I have found that keeping seeds in those cases results in yearly diminution of the quality of the germ plasm (reproductive cell DNA) as a result of inbreeding.

For some opponents, the foregoing consideration will be unsatisfactory because at the core of their opposition is a hankering after a new world economic order; agro-biotechnology is just one aspect in need of transformation. This would be a world in which commodities such as pharmaceuticals and seeds, to pick the two most frequently targeted, are not subject to intellectual property protection; a world in which research and development (R&D) is funded by governments, resulting in commodities that are 'owned' by the public, whose tax dollars presumably funded the R&D; they would be 'public goods'. Given the performance record of governments around the globe on systems such as this, I have little sympathy for this approach. More to the point, debates about the benefits, likelihood of success and so on of a new economic order miss a fundamental reality. We do not have such an economic order sitting in the wings – not even in communist China do we find such a model – and are unlikely to have one any time soon. Also, it is worth pointing out that alternative systems that have been implemented, from feudalism to Soviet communism, have significant deficiencies.

In the current economic order, the majority of medical, agricultural, environmental and consumer-product advances have been funded by private investors and undertaken by private companies. If their ability to realise an appropriate return on their investment and efforts is curtailed, a great deal of R&D on which we all depend will grind to a halt. This, of course, is not to say that there should be no public scrutiny and control over profits. Nor is it to say that where governments do make R&D investments (and most developed-nation governments do make such investments) the outcomes should not be public rather than private. Nor also is it to say that there are no challenges to be faced in the current economic order; there are many. To think, however, that these can or will be met by a new economic order in which public resources will fund R&D and the results are entirely a public good is at best naïve.

If someone's opposition to GM crops is not really specific to this biotechnology but is opposition to technology or corporations or capitalism in general, the parameters of the examination are very different from those relevant to an examination of opposition to this specific technology. Hence, a claim that the power that multinational corporations exercise in global and national affairs is harmful may be a legitimate topic for analysis and debate, but the analysis will be different, in numerous respects, from an analysis of whether GM crops that express *Bt* endotoxins are harming monarch butterflies. Obviously, it is very likely that any regulatory, legislative outcome of analysis and debate about the harms arising from the power of multinational corporations will have effects on companies involved in GM crop science and technology. Those effects, however, will not be a result of the fact they are a GM biotechnology company; it will be a result of the fact they are multinational companies. This is an important distinction. Frequently, individuals justify their opposition to GM crops by asserting that they are products of large multinational corporations. That, however, is not a justification for one's opposition to GM crops; it is a justification for one's opposition to a certain kind of corporation whatever its product.

For example, a multinational pharmaceutical company might produce an HIV/AIDS vaccine. To oppose the vaccine solely on the grounds that it is the product of a multinational corporation is confused reasoning. Whether the product itself is beneficial, harmful or suspect must be determined entirely independently of issues about the kind of company producing it. It is not inconceivable (indeed some might point to lots of past examples) that independent examination might conclude that the product is enormously effective with few side effects but the company is behaving irresponsibly in its production, pricing, distribution and so on. The immediate remedy might be regulatory intervention, legal action and the like, or, in really egregious cases, legislating approval for another corporation to produce the vaccine; the irrational (indeed ethically indefensible) remedy would be to ban production and use of the vaccine.

Notwithstanding the above analysis, there is one kind of harm that arises from multinational power and is of special relevance in the context of GM crops. It is still not a harm arising from GM technology or crops but rather from a feature of multinational power. Examples of this kind of harm include failure to allow access to GM crops to poor farmers in poor countries, or, the obverse, corporate enticements in poor countries to make inappropriate decisions with respect to crops, or the creation of dependence on a specific supplier, and the list goes on. These are legitimate concerns but they remain concerns about corporate power and behaviour and not specifically about harms arising from GM crops. Getting the actual locus of concern right allows a fruitful and relevant investigation, analysis, debate and action plan.

6.2 Environmental harms

Some purported environmental harms from GM agriculture are not unique to GM. Concerns, for example, about losing heritage stock, the development of resistance in pest populations, the effects of pesticide expression on non-target organisms, and the disruption of ecosystems resulting in a loss of biodiversity are common to all forms of agriculture. A few purported harms are more specific to GM agriculture; most centre on the introduction, into an environment, of a novel genetic organism. Hence, the potential for HGT or for the organism to become an invasive species poses unknown challenges. It is not that HGT does not occur with non-GM plants; it is that an 'artificial' gene complex might be spread to non-GM organisms. Similarly, examples of non-GM invasive species abound, but if a GM plant becomes invasive, an 'artificial' gene complex will begin to predominate with unknown consequences. Let's take the concerns one at a time.

6.2.1 Loss of heritage stock

The broad concern about losing heritage stock seems to be about loss of biodiversity. Even accepting, as I do, that biodiversity is important, this agricultural concern is too coarse-grained. Given the highly selected nature of agricultural animals and plants and the rarefied environments (quite artificial, human-created environments) in which they are raised and grown, it unlikely that their complete demise would have any important impacts on biodiversity. Indeed, quite the opposite seems likely; elimination of agriculture and a slow transformation of the environment from agriculture to forests, savannas, marshlands and the like will enhance biodiversity. So the debate about the dangers of losing heritage crops (there is much less concern raised about animals) is not really about loss of biodiversity but more about loss of agricultural plant varieties that existed 50 or more years ago. Nonetheless, this is a genuine concern; it is about losing varieties to which we may wish to return, either by growing them or using them in hybridisation to exploit genetic characteristics that seem relevant again.

This, however, is not an issue raised by biotechnology alone. Ordinary plant selection and hybridisation can lead and has led to loss of original varieties. That this is an issue that needs to be addressed emerged long before molecular biotechnology. In the 1950s, concern was expressed. Moreover, long before molecular biotechnology, solutions were put in place. Seed banking is now firmly entrenched in rich-country agriculture. A significant part of the reason this issue was addressed quickly and effectively is that the seed industry has a vested interest in not losing heritage varieties. Their science and technology depends on being able to return to heritage varieties to continue to develop plants with traits appropriate for changing conditions. Nothing will change with the entry of molecular biotechnology; that industry also needs to have access to past varieties.

One also must be clear about the scope of 'heritage' in this debate. If it refers to agricultural plants grown in the last 50–100 years, the concerns are about losing recent genetic traits that might once again have agricultural value. If it refers to some nostalgic hankering after 'natural' plants and animals, recapturing that world, assuming it is desirable, is well beyond reach. Consider potatoes, which originated high in the Andes of Bolivia and Peru. There are several thousand varieties but only 100 or so are used as food. Fewer than 10 varieties are commonly grown in rich countries. Humans have 'selected' from the variety of potatoes (the several thousand) those that have characteristics desirable for agriculture, food (taste, colour and nutrients), storage and handling. If natural processes were doing the selection, the probability that ones humans have selected would be favoured is vanishingly small. No organic farmer would take this human superseding of natural processes to be a reason to avoid growing the selected potatoes. If an organic farmer did, the acceptable plants and animals for farming would be near zero. The use of expressions such as 'heritage potato' or 'heritage tomato' is misleading. What we call a 'heritage tomato' bears little resemblance to any tomato variety 2,000 years ago – the real heritage tomatoes – or even to those found in Mexico by Cortes in 1519. As Harold McGee notes, 'Tomato started out as small berries growing on bushes in the west coast deserts of South America' (p. 329). The Aztecs, in Mexico, domesticated the tomato, which resulted in a fruit closer to ones familiar today. The real heritage tomato – one not manipulated by humans – is the original berry form. Humans, the Aztecs specifically, selected through hundreds of generations those that suited their purpose – not nature's purposes, not the tomato's purposes. Without human manipulation of nature, it is exceptionally unlikely (a probability approaching zero) that we would have anything remotely close to the contemporary tomato. The same is true of cattle, goats and other agricultural animals; they have all been domesticated by humans (shaped by artificial selection) for human purposes. The same applies to cattle, chickens and other farm animals. The 'natural' tomato, potato and cow bear little resemblance to what any kind of farmer today would want to grow or raise.

6.2.2 Horizontal gene transfer (HGT)

When genes are passed from parent to offspring, genes are transferred vertically. The genes in the offspring can be traced to genes in the parents. When genes pass to an organism from a non-parental organism, the genes have been transferred horizontally. The concern is that genes from engineered plants will transfer horizontally to other agricultural crops of the same variety, or, more worrying to some, to non-agricultural plants. It is important to be clear that HGT has occurred, and continues to occur, independently of GM, and, hence, only the genes inserted in a plant through GM are relevant. If a non-GM-inserted gene is horizontally transferred by natural processes, it may frustrate certain goals of humans, but, since it is a product of natural processes, it would occur even if GM had never been undertaken. Hence, only segments of DNA that have been inserted through GM and are transferred horizontally are relevant to an analysis of potential harms of GM.

At a superficial glance, this issue appears to pose a frightening prospect of genes inserted into GM plants finding their way through HGT into other plants, or even other organisms and taking over the landscape. Unfortunately, the molecular biology of HGT is complicated and those who sound alarm bells exploit this fact and provide few, if any, biological details. To sort out the reality from the bogus claims and impressions will require delving a little bit further into the molecular biology; the articles cited provide considerably more detail and more evidence than provided here, and they provide a wealth of additional citations.

HGT occurs frequently in prokaryotes (organisms without a nucleus: bacteria and Archaea are the two different domains of prokaryotes, and bacteria are well studied with respect to HGT). Nakamura and his colleagues (2004) found up to 25 per cent HGT in prokaryotes: 'We applied the Bayesian method[1] to analyze 116 prokaryotic complete genomes and found that 46,759 (~14%) of the total 324,653 open reading frames (ORFs) were derived from recent horizontal transfers . . . [for simplicity, open reading frames can be understood as any DNA segment that codes for a protein; the frame is the sequence, which is read in the process of constructing the protein]. The average proportion of horizontally transferred genes per genome was ~12% of all ORFs, ranging from 0.5% to 25% depending on prokaryotic lineage' (p. 760). As Richardson and Palmer (2007) note, however, 'With very rare exception, HGT occurs much less frequently in eukaryotes than in bacteria, although the process may have been more common early in eukaryotic evolution' (p. 1). Moreover, in eukaryotes, HGT of nuclear DNA is exceptionally rare. Richardson and Palmer found, 'Nuclear HGT is rare in multicellular eukaryotes (animals, fungi, and plants). Nearly all known cases involve bacteria as donors' (p. 1). They also found, 'Among the plants, *Agrobacterium rhizogenes* has donated genes, some functional, to members of its host genus *Nicotiana*. During pathogenesis, *Agrobacterium* transforms its host with several plasmid-encoded genes, with HGT as a natural consequence. Additional putative cases of bacterium-to-plant nuclear genome HGT (outside of organelle-to-nucleus IGT [intracellular gene transfer]) include the

[1] An analytical method based on repetitive sequential application of Bayes' theorem: $Pr(H/E) = [Pr(H)Pr(E/H)]/[Pr(H)Pr(E/H) + Pr(\sim H)Pr(E/\sim H)]$, where H is a hypothesis, E is some evidence, and $Pr(\Phi/\Theta)$ is the probability of Φ given Θ.

acquisition of aquaglyceroporins from a eubacterium; 1200 million years ago and of glutathione biosynthesis genes from an alpha-proteobacterium' (p. 2).

Most HGT in eukaryotes involves mitochondrial DNA: 'The low levels of horizontal transfer of nuclear genes in multicellular eukaryotes contrasts with evidence that their nuclear transposable elements have moved horizontally on numerous occasions, although relatively few such transfers have yet been documented in plants. Like nuclear genomes and yeast mitochondrial genomes, plant mitochondria have been subject to horizontal transfer of mobile genetic elements. Most notably, the discovery of high frequency angiosperm-to-angiosperm horizontal transfer of a homing group I intron [introns are segments of DNA that do not code for a protein and exons are segments that do code for a protein; when messenger RNA (mRNA) is transcribed from DNA, introns are cleaved and removed] in the mitochondrial cox1 gene (Cho *et al.*, 1998; Cho and Palmer, 1999) foreshadowed the recent discovery of widespread horizontal transfer of plant mitochondrial genes' (Richardson and Palmer, 2007, p. 2). Two Cox genes (cytochrome oxidase genes) have been identified in humans: Cox-1 and Cox-2. Cox genes are mostly found on the nuclear membrane but are also found in mitochondria. For more on this, see Chandrasekharan and Simmons (2004).

The fact that all known cases of nuclear HGT involve bacteria as donors raises the issue of bacteria as intermediaries in nuclear HGT; that is, from plant to bacterium to plant. The first step in such a sequence is plant to bacterium HGT. This issue has recently been examined by Brigulla and Wackernagel (2010):

> Considering the occurrence of HGT processes in nature, the question arose on the likelihood and the possible frequency of HGT from GMOs to prokaryotes. Eukaryotic DNA can transit to prokaryotes only through transformation. With respect to tg [transgenic] plants, transformation would require the release of DNA from plants (during growth, wounding of tissue, pathogen infection, pollen spread, or death), persistence of the DNA in the environment (e.g., in soil, water, plant rhizosphere, intestinal tract, etc.), interaction of DNA with competent prokaryotic cells, and finally, genomic integration of DNA in the new host. (p. 1033)

We will look more carefully in a moment at the impediments to HGT in eukaryotes and especially in the case of GM plants. We will also look at the 'background DNA-sequence noise' that swamps any possible HGT of a sequence via GM plants. For now, the stage for this is set by the concluding remarks of Brigulla and Wackernagel (2010), the details of which will be more comprehensible after further elaboration of HGT:

> The possible transfer of genes from GMOs to prokaryotes in their natural habitats has been addressed. A survey of 60 genes presently introduced into tg plants for agricultural and industrial purposes indicates that the genes come mostly from prokaryotic genomes, but also from eukaryotes, prokaryotic genetic elements, and viruses. The genes are naturally abundant in prokaryotic habitats. In the tg [transgenic: GM] organisms, the genes are present unaltered or with modified or recombined nucleotide sequence. If a gene is foreign for a prokaryotic recipient and not embedded in homologous sequences [see p. 24 above], experimental data indicate that under natural conditions, the chance for a successful transfer would be about 7×10^{-23} per cell or less. [10^{-x} is a compact notation for 1 divided by 1 followed by x zeros. For example, $10^{-3} = 1/1{,}000$. So, if the chance for a successful transfer would be about 7×10^{-23} per cell or less, that means the chance for a successful transfer would be about 1/700,000,000,000,000,000,000,000 or less, making the chance vanishingly small.] This frequency is much lower than the gene transfer frequency among prokaryotes observed in the environment (about 10^{-1} to 10^{-8} per cell). There is a wide and continuous movement of genes among prokaryotes via HGT in the natural environment which is not limited to genetic material from prokaryotes. The mobilome constitutes a central element in this phenomenon. The homologous and IR [illegitimate recombination = recombination mechanisms that do not involve homologous regions of DNA; Homologous recombination is discussed in Section 2.1] mechanisms for integration of DNA and for the creation of new arrangements of genes and parts of genes allow one to explore by HGT the effect of any combination of genetic material on the fitness of prokaryotic cells. Compared to this continuing natural experimentation in the biosphere, the introduced genetic constructs in GMOs appear marginal. Nevertheless, it is prudent and responsible to survey, as it is standard practice worldwide, any construct for possible negative effects on human and animal health and the ecosystem before release of the GMO and to observe the GMO after release. (p. 1037)

The infrequency of HGT in eukaryotes, especially nuclear HGT, is not surprising. The complex conditions that must be satisfied for HGT to occur are extensive. Four different, and critical, aspects can be identified. First, the transfer has to occur; that is, the segment of DNA has to be physically transferred from one cell to another. Second, the transferred DNA has to be integrated into the nuclear or mitochondrial DNA of the receiving cell. Third, integration of the segment of DNA must not destabilise the cell, rendering it non-viable. Fourth, a viable cell with the segment of DNA incorporated must have a sufficient selective advantage to result in an increase in its proportion in the relevant population. Let's look at them one at time.

If integration occurs, it is only successful – 'successful' meaning not only that the host is viable but also that the integrated segment is expressed – if several other conditions are met (see Brigulla and Wackernagel, 2010). Clearly, any inactivation or deletion of essential genomic functions that results from the integration will render the cell(s) non-viable. This restricts the location of integration to regions of the genome where there are non-essential or redundant sequences, or between essential sequences (intergenic regions). Moreover, the integrated genes cannot be toxic to the host; that is, they cannot code for the production of a protein that is toxic to the host or that will increase the production of a protein to levels that are toxic.

A further impediment to successful integration is silencing. Cells employ several mechanisms to protect the integrity of their DNA. One, set out in Section 2.1, is the restriction-modification system, which produces restriction enzymes that cleave foreign DNA at its restriction site – hence destroying it. Recall, a cell stops, through a methylation system, its own restriction enzymes from damaging its DNA. Silencing is another protective mechanism. A histone-like nucleotide structuring protein (H-NS) binds sequences with diverging G–C base pairings. Brigulla and Wackernagel (2010) point to research suggesting two purposes of silencing, 'The silencing by transcription downregulation may, on one hand, act as a protective mechanism against invading foreign DNA (Navarre *et al.*, 2007) and, on the other hand, may support gene retention for the further functional integration into interaction networks by mutations' (Lercher and Pal, 2008, p. 1033). In bacteria, there are widespread mechanisms for counteracting this silencing, which is part of the explanation for the prevalence of HGT in bacteria. These counteracting mechanisms are much less common in plants and animals; hence, silencing much more frequently mitigates HGT.

So, to this point, the requirements for successful HGT are: (1) the foreign gene segment has to be physically transferred to the host, (2) the foreign DNA must become integrated into the host DNA (this usually requires the existence of a homologous region of DNA and cleavage at the relevant site), (3) the foreign DNA must not be inserted in a region that inactivates or deletes essential genomic functions, (4) the foreign DNA must not code for a protein toxic to the host, and (5) the foreign DNA must elude the host silencing mechanism. Not surprisingly, given these requirements, successful HGT is very rare in plants and animals. There is, however, yet another requirement. The effect of the foreign DNA, should it meet requirements (1)–(5), is that it does not reduce the fitness of the phenotype. If it reduces the fitness of the phenotype, it will be eliminated by selection. In the overwhelming majority of cases of HGT in plants, the changes to the phenotypes make them less fit (i.e. the changes are deleterious), and that phenotype will be eliminated by selection.

The cumulative effect of these requirements suggests that although the probability of HGT of a DNA sequence that has been inserted into a GM plant is not zero, it is highly unlikely. Add to this the fact that the segments engineered into plants are already available in abundance in nature in non-GM organisms, and the probability of a GM-caused harmful HGT is vanishingly small. Richardson and Palmer (2007) encapsulate this well with respect to *Bt* Cry toxins:

> A large group of prokaryotic genes in tg plants codes for insecticidal Cry protein variants from *Bacillus thuringiensis* strains. *B. thuringiensis* strains with genes for Cry toxin production are present in a great variety of habitats. Strains have been isolated from the gut of insects as well as from soil and the surface of plants including fresh fruits and vegetables. The analysis of human nasal swab samples indicated a wide distribution of *B. thuringiensis* in the human population. Thus, in general, the genes integrated in tg plants are naturally abundant in the prokaryotic and eukaryotic world. A HGT of these genes could have occurred for extensive periods from the original source to a new host. (p. 1036: citations provided in the original have been removed)

In connection with this point, it is worth highlighting that, as noted above, nuclear HGT in plants is extremely rare. Significantly, in those cases where it does occur, nearly all involve bacteria as donors. The DNA segments inserted into GM plants are widespread in the bacterial world and cases of bacteria to plant nuclear HGT have been observed. Hence, the probability of naturally occurring bacteria-to-plant nuclear HGT of the same DNA segments as those inserted in GM plants is very much higher than any plant-to-plant nuclear HGT, the latter being extremely rare to non-existent.

A key point in the above examination is that mitochondrial HGT is much more common than nuclear HGT in plants, as Richardson and Palmer (2007), quoted above, have demonstrated. Mitochondrial DNA is extra-chromosomal DNA (outside the nucleus); it is found in organelles in cells that are known as mitochondria. Unlike nuclear DNA, which determines, through a complicated developmental process, the traits of the phenotype, mitochondrial DNA is only involved in the metabolic processes of cells; it is mostly involved in the metabolic process of converting food into energy within the cell. There is considerably more intracellular mitochondrial gene transfer than mitochondrial HGT. GM plants have the relevant DNA segment inserted into nuclear DNA, since the goal is to have agriculturally beneficial traits expressed in the phenotype; hence, any HGT related to GM plants will not be mitochondrial HGT but rather nuclear HGT, which, as set out in some detail, is exceptionally rare.

The claim that the cumulative effect of our current knowledge of HGT suggests that GM plants pose no new risks is consistent with empirical evidence; no HGT of DNA sequences inserted in GM plants has been observed. This is no reason for complacency or unbridled enthusiasm about GM. Norman Ellstrand (2001), in an article on the potential for hybrids of GM and 'wild' plants to occur and create problems (a potential form of HGT), captures the prudent stance well:

> The products of plant improvement [non-GM improvements] are not absolutely safe, and we cannot expect transgenic crops to be absolutely safe either. Recognition of that fact suggests that creating something just because we are now able to do so is an inadequate reason for embracing a new technology. If we have advanced tools for creating novel agricultural products, we should use the advanced knowledge from ecology and population genetics as well as social sciences and humanities to make mindful choices about how to create the products that are best for humans and the environment. (p. 1545)

This strikes again the chord that balancing benefits and harms is essential and doing so requires a broad set of considerations.

6.2.3 Development of resistance

One challenge – probably the only one – that *Bt* shares with all other pesticides is the potential for resistance to evolve in the target organism. The development of resistance is a function of selection acting on the high level of variability (genetic and phenotypic) in populations of organisms. Resistance to the toxin may exist in some organisms in the population, or a mutation may arise which confers resistance on the mutant. Resistance is usually dose dependent such that large quantities of the toxin applied over an extended period will affect all but the most resistant organisms. Leaving aside GM crops with the *Bt* endotoxin expression for a moment, an application of *Bt* spray in an organic field will kill all the target organisms unless there are organisms resistant to that level of the toxin. If there are resistant organisms, they will be the only ones left to reproduce. If only resistant organisms survive the *Bt* spraying, there will be complete resistance to that dosage after a single generation. In actual cases of the development of resistance, it takes several (but not very many) generations to fully develop; this occurs for a variety of reasons such as differential spatial density of the sprayed toxin (some areas will receive more than others due to factors such as wind-caused drifting) and the scalar nature of resistance (some organisms will have minimal resistance to the toxin, others a high resistance, i.e. tolerance).

Clearly, it is not in anyone's interest – except for the target organisms – to have resistance develop to an initially effective pesticide, especially not to a pesticide that after 50 years of use has no known negative environmental, wildlife or human health impacts. Two key biological elements used to delay the development of resistance are toxicity dynamics and population genetic dynamics. Research to determine the toxicity level of different levels of exposure is essential. *Bt* ideally needs to be applied at rates (quantitatively and temporally) sufficient to kill all of the target organisms. Unfortunately, even with excellent toxicological data, application can be uneven and, more importantly, mutants can develop that have resistance to the prescribed application rates. GM crops engineered to express the *Bt* δ-endotoxin resolve the consistency of application issue because all the plants express the endotoxin. To resolve the other causes of potential resistance, toxicological approaches have been supplemented by population genetic approaches.

Although population genetic modelling is complex, involving many variables, the essence of this method of delaying the development of resistance can be set out quite simply. Farmers are required to plant a refuge crop (a non-GM crop); the refuge crop must be 20 per cent of the total hectares planted and arranged in one of the prescribed patterns (Figure 6.1).

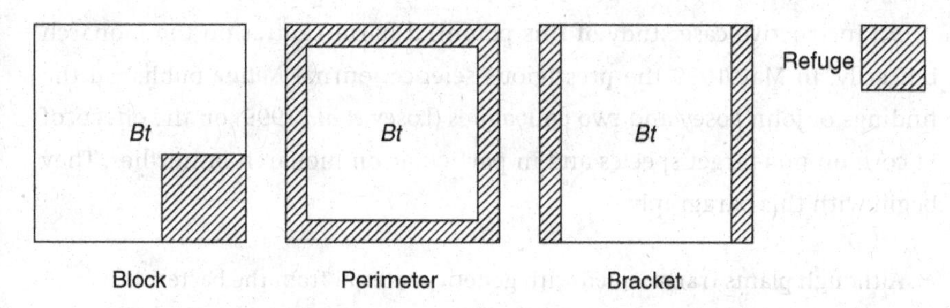

Figure 6.1 Examples of refuge-planting patterns.

This ensures that any larvae that survive the *Bt* δ-endotoxin will breed with a large population of larvae not exposed to the *Bt* δ-endotoxin. The most widely used GM crops with the *Bt* δ-endotoxin expression are maize and cotton; tomatoes and potatoes have been engineered and approved but are not widely used. Bourguet *et al.* (2005) provide an excellent account of insect resistance management (see also Ambec, 2005; Crespo *et al.*, 2009; EPA, 2001). Since management measures only delay, significantly in this case, resistance, GM companies also use stacking (multiple traits in a specific crop) to further protect against the development of resistance.

6.2.4 Effects on non-target organisms

The targets of agricultural pesticides are organisms that decrease the yield or quality of a crop. A challenge is to minimise the impact on other organisms – those that do not decrease yield or quality: non-targeted organisms. A pesticide, or any other agricultural chemical for that matter, that is toxic to fish and can be expected to make its way into rivers and lakes affects non-target organisms – fish. Ideally, a pesticide will have no effects on non-target organisms but this is seldom the case. One pesticide that comes very close to this ideal is *Bt* δ-endotoxin, as already described above. A number of GM crops contain the gene for expressing this endotoxin. Even though *Bt* has been used as a pesticide spray for many decades, concern about GM *Bt* crops focuses on the potential effect its production by a plant – instead of the bacterium – might have on non-target organisms. For example, the toxin is expressed in the pollen of the plant, as well as in other locations, and the pollen of a plant such as corn can be carried to distant locations by wind.

An instructive case study of this potential harm centres on the monarch butterfly. In May 1999 the prestigious science journal *Nature* published the findings of John Losey and two colleagues (Losey *et al.*, 1999) on the effects of *Bt* corn on non-target species and in particular on monarch butterflies. They begin with this paragraph:

> Although plants transformed with genetic material from the bacterium *Bacillus thuringiensis* (*Bt*) are generally thought to have negligible impact on non-target organisms, *Bt* corn plants might represent a risk because most hybrids express the *Bt* toxin in pollen, and corn pollen is dispersed over at least 60 metres by wind. Corn pollen is deposited on other plants near corn fields and can be ingested by the non-target organisms that consume these plants. In a laboratory assay we found that larvae of the monarch butterfly, *Danaus plexippus*, reared on milkweed leaves dusted with pollen from *Bt* corn, ate less, grew more slowly and suffered higher mortality than larvae reared on leaves dusted with untransformed corn pollen or on leaves without pollen. (p. 214)

It is important to note that this was a laboratory study in which exposure to *Bt* corn pollen was at levels much higher than would occur in a natural setting. Nonetheless, it raised the potential for a significant negative effect on a non-target organism. In addition, the monarch butterfly could reasonably be seen as the canary in the mine. Their results were published in 1999 – four years after GM crops had been approved for agricultural use – and the research occurred earlier. These were early days of GM agriculture, and due diligence and vigilance required this kind of work be undertaken and the results taken seriously, as they indeed were. Research teams in Canada and the USA began to investigate the impact of *Bt* corn on monarch butterflies in natural environments, and both the Canadian Food Inspection Agency and the EPA in the USA began to examine their approval processes and regulatory oversight. Mark Sears, at the University of Guelph in Ontario, Canada, and seven international colleagues published the results of further and extensive research in 2001 in *Proceedings of the National Academy of Sciences of the USA*. As the abstract indicates, this was multi-university collaboration: 'A collaborative research effort by scientists in several states and in Canada has produced information to develop a formal risk assessment of the impact of Bt corn on monarch butterfly (*Danaus plexippus*) populations' (p. 11937). Their research, corroborated by others, indicated that *Bt* corn has a negligible effect on monarch butterfly populations. *National Geographic News* (3 May 2010) reported, 'Sears pointed out that he has witnessed more damage to the butterfly population through "road kill" while driving along country roads than he did in his experiments' (http://news.nationalgeographic.com/news/2001/09/0910_wiremonarchs.html).

The initial study, the controversy it sparked, and the regulatory discussions and further research it kindled are a model of how novel technologies should be monitored. It has now been more than 15 years since GM crops were planted and a wealth of research has been undertaken. No documented cases of negative effects on non-target organisms have been found, the hype of critics notwithstanding. The absence of instances to date obviously does not mean that there are no negative effects on non-target species, but the probability that there are is diminishing by the year and with each new research report – see the meta-analysis by Marvier *et al.* (2007).

6.2.5 Disrupting ecosystems

There can be no doubt that agricultural practices have for millennia been transforming ecosystems. Clearing forests, scrubland and grassland and putting the land into agricultural use has a dramatic effect on ecosystems, and, as we are learning somewhat too late, on biodiversity. Managing weeds and pests, which all farmers must do in some way, affects the natural ecosystem dynamics. Even an organic farmer who removes Colorado potato beetles from her potato crop is removing from other organisms a potential food source and, thereby, affecting their survival. Few people argue that we should stop producing food and return to hunter-gatherer times, and most people recognise that controlling weeds and pests in some way is essential. So, the relevant question is, 'Does biotechnology result in greater (or lesser) ecological disruption than other agricultural practices?' As set out in Section 5.1 above, the evidence suggests it results in less ecological disruption. In Section 7.1 below, I focus specifically on the claims made about organic agriculture and environmental/ecological impacts. The case developed there is that organic farming has less impact than conventional farming but not by as much as many claim; compared, however, to biotechnologically based agriculture, the two are, at best, tied, and there are solid reasons for believing biotechnologically based agriculture already has less impact and in the near future will have significantly less.

One point worthy of note in this context is that ecosystems in which agriculture is a central feature are already artificial in the sense that the introduction of agriculture into that environment dramatically changed the previously 'wild' ecosystem. Moreover, the removal of established agriculture from an environment would involve an equally dramatic change, a change to a new 'wild' ecosystem – but not the one originally transformed. Further, wild ecosystems are, themselves, not static; they are dynamic, ever changing, systems in which new 'forms most beautiful' are evolving and other forms going extinct. When agriculture, any kind of agriculture, is introduced into an area, it disrupts the existing ecosystem. Land is cleared of trees, shrubs or grasses and is planted with agricultural crops or prepared for animal grazing or planted with plants for animal fodder and bedding (straw, for example). The ripple effect on existing plants and organisms is rapid. Sometimes the changes are beneficial to groups of organisms, sometimes detrimental. Very quickly, a new ecosystem develops. There is no *a priori* reason to prefer the original ecosystem and species to the new ecosystem. Hence, those who lament the changes that some new agricultural practice may introduce need to articulate reasons why the old is preferable to the new. Simply resisting change in the biological world is inadequate; change is the rule in the natural biological world, stasis the exception. And, to reiterate, the introduction of agriculture, of any kind, into an environment is the most dramatic disruption; adjustments to new agricultural practices thereafter most often have, at most, modest effects. However, there are two concerns that are not a simple resistance to change.

One concern is not that this or that species may diminish in number in an area or be entirely eliminated in that particular area but that a rapid pace of change over a large area may destabilise an ecosystem such that ecosystem collapse is inevitable. The collapse results in the permanent loss of those species, and if the area encompasses all the members of a species, extinction occurs. That this is happening as rainforests are cleared for agriculture, in, for example, Brazil, is indisputable; that the rapidity and scale of the loss of biodiversity as a result is alarming is also indisputable. There are many reasons to be alarmed but the one that resonates with people, because of the obvious human self-interest involved, is the discovery of new medicines. The renowned Harvard evolutionary biologist E. O. Wilson has captured this succinctly: 'It is no exaggeration to say that the search for natural medicinals is a race between science and extinction, and will become critically so as more forests fall and coral reefs bleach out and disintegrate' (Wilson, 1975, p. 123).

This loss of biodiversity is indeed a cause for alarm. Even though ecosystems change and this naturally results in the loss of species here and there (sometimes through extinction, sometimes through evolution), the rapidity and scale of the human-caused loss is reason for extreme concern. There can be no doubt that agriculture has had these alarming consequences, but most of the rapid and broad loss of biodiversity has resulted and will continue to result from the initial introduction of agriculture into an area. That is not to suggest that certain changes in practices in established agricultural areas cannot have negative effects on species survival – it is just that they are seldom as rapid or broad. That, one might reasonably claim, makes them more insidious and less likely to be attended to until disaster is at the doorstep, and this takes us directly to the second concern.

This concern is that changes in an agricultural area will spill over to non-agricultural areas. One example is the pollution of rivers and lakes that affects organisms hundreds of kilometres downstream. The pollutant might be synthetic fertilisers, herbicides and pesticides or organic materials from animal feedlots or manure deposition, storage or use. It is important not to focus just on synthetic products. The lessons of the last 50 years are that conventional and organic agriculture produce pollutants and, hence, can potentially spread disaster beyond the agricultural boundaries. In some cases, organic pollutants have been devastating, even to human populations whose drinking water has been contaminated with potentially deadly *E. coli* from manure runoff. A tragic and dramatic example of this occurred in 2000 in Walkerton, Ontario, Canada. During a period of heavy rain, the exceptionally dangerous O157:H7 strain of *E. coli* leached into the groundwater from the surrounding animal agricultural operations. This contaminated the community's water supply. Several thousand people got sick and at least seven people died as a result. Human failings were a clear contributing factor – inadequate chlorination and regulatory failings, for example – but the source of the problem was cattle agriculture. Other organisms do not chlorinate or otherwise purify the water they drink; they just drink it and suffer the adverse effects. A less dramatic example is the *E. coli* contamination of California spinach in 2006. Meat, vegetables and water, at various times, have been contaminated with cattle-source, extremely dangerous *E. coli*.

Consider also nitrogen added to a field, whether from manure or synthetic urea. It will leach into the water table and from there into rivers and lakes from where it will enter the atmosphere as a greenhouse gas. That effect is widespread, changing simultaneously numerous ecosystems. Hence, there is no cause for complacency just because the effects of an agricultural practice are slow and multifaceted.

That there are troubling, indeed with respect to some facets, alarming effects of agriculture on ecosystems and, in turn, biodiversity is difficult to deny. There is some uncertainty about some aspects, but, on the whole, the evidence suggests that agriculture is a major component in ecosystem degradation. In light of this, two central questions are: what can be done to mitigate these negative effects and are there grounds for singling out biotechnology as a special villain in the degradation of ecosystems? The second question was addressed in Section 5.1 and will be examined again in Section 7.1. I argue in both places that agricultural biotechnology is not a special villain. Indeed, I argue, there are reasons to believe that biotechnology is our only effective way forward in both producing enough affordable food and reversing environmental degradation.

The first, and more general question, does not admit of simple answers. With world population at 6.8 billion and growing, and a population of nearly 1 billion in Europe, North America, Australia and New Zealand combined, eliminating agriculture is not an option; even reducing agricultural output is not an option. What might constitute an important first step is prohibition of creating new agricultural land. This, in rich countries, has gained considerable support but has been met with considerable cynicism in less affluent nations. A Brazilian member of a committee, on which I also served, passionately exclaimed during one meeting, 'It's pretty hypocritical of North Americans to be campaigning to save the rainforests and chastising Brazil for allowing their destruction so as to increase our agricultural output when North Americans have already denuded their landscape of forests and enjoy the agricultural benefits of doing so.' He was well aware that destruction of rainforests was courting future disaster, but his point was that sanctimonious preaching by those who have long ago done just what Brazilians are now being told is wrong has a hollow ring. Nonetheless, halting the relentless destruction of forests, scrublands and grasslands to increase the agricultural land base is essential. Any arrangement to do so, however, must take care to distribute the benefits and harms equitably. For North Americans and Europeans to retain their agricultural base and the abundant and affordable food it produces while denying those in low- and middle-income countries the chance to 'catch up' fails to deliver an equitable distribution. An equitable distribution can be achieved in many ways; increasing yields on existing land, for example, and providing food to other countries as compensation for not creating more agricultural land (see Kalder–Hicks compensation principle in Section 3.3 above) or providing a straightforward monetary compensation for not creating more agricultural land. Whatever arrangements are put in place, the result cannot be that some groups bear most of the harms while other groups enjoy most of the benefits; that would violate a fundamental principle of justice, fairness and equality.

One thing is clear; arresting ecological and environmental degradation due to agriculture depends minimally on three things: stabilising the current agricultural footprint (or, to the extent possible, reducing it), decreasing the use of pesticides, herbicides and synthetic fertilisers, and decreasing the impact of animal agriculture. It is difficult to imagine how these can be achieved without a contribution from agricultural biotechnology. Perhaps if everyone in the world became vegan, these could be achieved but this is simply not tenable. Encouraging people to consume less meat, dairy products and eggs should undoubtedly be a component in a viable strategy, but any strategy that depends on a massive and dramatic change in dietary behaviour is pure fantasy and is doomed to failure. Following the example of John Adam (2009), we can do a simple calculation of meat demand in rich countries. The combined population of Australia, Europe (including the UK) and the Americas, in 2010, is approximately 1.4 billion. Setting assumptions on the conservative side, suppose 50 per cent are vegetarians or infants and the other 50 per cent eat 0.25 kg of meat a week (=0.0357 kg per day). Then the overall average daily consumption will be 1.4 billion ×0.5 (=700 million) × 0.0357 = 25 million kg. One shorthorn cow weighs about 908 kg. So daily consumption of meat in rich countries is equal to about 27,500 shorthorn cows per day. This is an equivalence since shorthorns will not comprise the entire meat-consumption profile. The sources of meat are varied, but the vast majority of meat comes from cattle, pigs, sheep, goats and chicken/turkey (it takes a lot of chickens to equal one cow). The total shorthorn-equivalent consumption for a year is about 10 million shorthorns. This is a low-end estimate since more than 50 per cent of people in rich countries eat meat and those that do, consume more than 0.25 kg a week on average over a year. For example, 0.25 kg per week (raw weight, which is what the butchered animal yields) equals two Macdonald's quarter-pound hamburgers per week or one-half a chicken breast per week. Add dairy cows and egg-laying chickens, and the barnyard animal numbers required climb even higher. Hence, in Australia, Europe and North America (major rich countries), at a minimum, the meat equivalent of 10 million shorthorn cows is consumed per year. As shown in Chapter 7, organic agriculture cannot supplant conventional agriculture with demand for crops and animals at current levels; even with a dramatic reduction in demand, which is not going to occur, organic could not alone supplant conventional agriculture and its environmentally destructive practices. Organic and GM combined have a fighting chance of doing so.

6.3 Health harms

The human health concerns seem less significant now that we have 15 years of experience with widespread consumer consumption in North America (population of USA and Canada: about 350 million) of products containing GM plant material (e.g. soybean oil and canola oil); **no new health risks have emerged.** Of course, one must, nonetheless, be constantly monitoring and researching the issue, as with any product – even conventional foods such as meat and dairy products. In the next chapter, I examine in considerable detail the debate about health risks. The examination occurs in the comparative context of GM, organic and conventional agriculture and provides, thereby a more grounded discussion.

One thing worthy of note here is the remarkable difference in public perception of GM medicine and GM agriculture, a point I have raised before and which will arise again in Chapter 8. GM is thriving in four main areas: agriculture, medicine, environmental remediation and aquaculture.[2] Of these, crop agriculture has received the most negative public, media and policy attention, with aquaculture episodically receiving such attention. On the surface, this is surprising since GM in crop agriculture pales by comparison with the current use of GM in medicine and the significant array of GM research currently

[2] There has been considerable media attention paid to molecular genetics in forensics (paternity issues, rape and murder cases, etc.). For the most part, these do not involve modification of the DNA of an organism. They involve molecular techniques such as polymerase chain reaction (PCR) – a technique which allows small samples of DNA to be amplified. The goal is not the creation of new traits in organisms. Industrial applications of molecular biotechnology have a long history – the use of enzymes and yeast in the food and beverage industries, for example – but recombinant techniques do not play a significant role.

under way. Among the numerous current applications of GM in medicine and those that are the focus of intense research are gene therapy, creating monoclonal antibodies, pharmacogenomics, artificial blood, tissue engineering, xenotransplantation, stem cell-based therapies and therapeutic cloning. GM in medicine has a longer history than in agriculture and arguably, the potential harms are much greater with GM in medicine. For example, strains of *E. coli* reside in the human intestinal tract and are an essential component in digestion. These are, obviously, non-pathogenic, but pathogenic strains are common and produce pyogenic infections and diarrhoea (these include enteropathogenic strains and enterotoxigenic species). Methods of ensuring that GM *E. coli* bacteria remain confined to the laboratory or cannot survive outside a narrow range of laboratory conditions have to this point been successful. But, bacteria mutate quickly, so the probability is not zero that a GM strain that is pathogenic to humans could mutate and survive outside the laboratory.

The point is not that dire hazards are on the horizon – they are not – but that GM in medicine poses risks as well as benefits just as GM in agriculture poses risks as well as benefits. The question this raises is, 'why has GM in agriculture come under substantially more public and political scrutiny than GM in medicine?' Pointing out that GM in agriculture rests in the hands of large multinational private enterprises does not seem a fruitful explanatory tact. Producing pharmaceuticals, such as recombinant insulin to treat diabetes and factor VII to treat haemophilia, involves genetically modifying organisms (bacteria), and there are many other recombinant therapeutic proteins manufactured with GM bacteria: DNase (cystic fibrosis), erythropoietin (anaemia), granulocyte colony-stimulating factor (white blood cell deficiency), interferons and interleukins (leukaemia), superoxide dismutase (tissue damage from heart attack), tissue plasminogen activator (heart attack and stroke) and a variety of vaccines. In addition, there are many other therapeutic agents and techniques that use GM. Pharmaceutical companies, which are the producers of many of these therapeutic agents, are also large, multinational private enterprises; indeed, they make companies like Monsanto and Syngenta seem small. Sporadic intense media attention has been focused on pharmaceutical companies, but most has to do with ethical issues around marketing practices, distribution decisions, suppression of negative trial results, influence peddling and the like; very little attention has been focused on GM practices.

The difference in the public and political treatment of agriculture and medicine is, I think, multifaceted. Part of the explanation centres on differences in the nature of the benefits, part on differences in trade implications, part on the mystical and primordial human relationship with food, part on differences in the accessibility of information, and the list goes on. Though just one element in a full explanation, inaccessibility of information should not be underestimated. Overall, the scientific knowledge of the public and of politicians is impoverished. Crafting a simple intelligible message for public consumption (to galvanise opinion or concern, for example) is a daunting task. For medicine, the task is amplified by the clinical application of the science. Most people (the public, politicians and media workers) accept their limitations; a few, imprudently, occasionally put their ignorance on public display. Not so with food; expertise abounds and most people have an unsupportable confidence in their ability to comprehend the food process from farm to table. Even a modest foray into the literature on food science, food economics, nutritional science and food manufacturing suggests that 'overconfidence' best describes the public's sense of knowledge of food.

It is likely that another important difference between medicine and agriculture is that in rich countries GM in medicine matters, a point touched on again in Chapter 8. Advances in science, technology and their clinical medical applications save lives and ameliorate compromises to health. A ban on GM research and applications in medicine will mean increased mortality and morbidity as numerous therapeutic agents and techniques cease to be available; not to mention that future decreases in mortality and morbidity will be compromised. In rich countries, a ban on GM research and applications in agriculture will make little difference. Food is plentiful and inexpensive and regulatory processes have made it safer than at any time in human history. GM crops may be terrific for farmers in terms of lower input and labour costs and increased yields, but even without GM crops, starvation, malnutrition and the like will not increase in rich nations. Hence, people in rich nations have the luxury, for a few more decades at least, of lofty anti-GM rhetoric on food but not on medicine. This might strike some as blatant hypocrisy. Hypocrisy involves intentional acceptance and/or promotion of two contradictory positions. That language is appropriate when governments and NGOs accept, indeed exploit, contradictory positions. Things are more complicated at the individual level. There are certainly some people who know that the positions they are advocating on GM are inconsistent. They know their positions on activities in medicine and agriculture are unjustifiably different and they deliberately deceive their audiences to advance their own agendas. That, at the individual level, is hypocrisy. For many people, however, the inconsistency is pre-analytic; they have never reflected on the issues in a cross-boundary way and, hence, have not detected the inconsistencies. Once their attention is drawn to the inconsistencies they become uncomfortable with their commitment to such inconsistent views. Those who do not take action to resolve the problem might be lazy or indifferent but most are likely unsure about how to proceed. They accept that their views, as a whole, are irrational but are stymied with respect to a resolution. A goal of this book is to provide some tools for detecting inconsistencies (such as the foregoing) and resolving them.

7 The organic alternative

This chapter examines the purported benefits of organic agriculture – environmental and health benefits – to avoid confusion, it must be noted, at the outset, that the term 'organic' in chemistry has a different meaning than in agriculture; organic chemicals are those containing carbon. I use the expression 'organic agriculture' rather than 'organic farming'. This is deliberate; 'organic farming' conjures up for many the image of the family farm – a small operation with a few cows, goats, sheep, pigs and chickens, all of which roam free in the pasture, and a vegetable plot managed by family members. There are, of course, still a few such operations, but organic agriculture has become big business; it is an agricultural business in every sense that conventional agriculture is. It is because of this that nearly every large and small grocery store across North America and Europe has been able to supply customers, daily, with an array of organic products. Moreover, the rapid growth and current magnitude of organic agriculture has warranted the commitment of time and financial resources to formulate policies and regulations and to engage in inspection and enforcement.

In what follows, I separate GM agriculture from conventional agriculture; advocates of organic agriculture lump them together, portraying them both as the antithesis of organic. This, I argue, results in a crude analysis that undermines creative approaches to meeting the serious challenges facing agriculture, health promotion and environmental remediation. As we shall see, much of the attractiveness of the case mounted in support of the benefits of organic agriculture rests on 'intuition' and anecdotes. Since organic agriculture avoids chemicals and 'factory farming', it seems **intuitively** obvious that this must reduce environmental impact and produce healthier food; after all, having no chemicals means less environmental pollution – and, hence, more sustainable farming – and eliminating all those chemical residues from food will, obviously, make them less damaging to one's health. Intuitions, however, as we have seen and will see again here, are a poor substitute for empirical evidence; this is a message that has been reinforced in contexts as diverse as medical practice and military strategy over the last several hundred years.

Empirical evidence does support some of the claimed environmental benefits but it is far less compelling with respect to the health benefits. In addition, where the evidence supports some environmental benefits, the comparison is between non-GM conventional agriculture and organic. That, however, is not the most helpful comparison since when GM agriculture is compared directly with non-GM conventional agriculture, it also wins, and by a large margin. What is environmentally unsustainable is non-GM conventional agriculture. Both organic and GM agriculture contribute to a reduction of the environmental degradation associated with non-GM conventional agriculture. The claims of Greenpeace[1] and its sympathising NGOs aside, there is an emerging view that the dichotomy between GM and organic agriculture is overstated and detrimental to both (see Ammann, 2008, 2009; Ronald and Adamchak, 2008). In addition, even if, contrary to the evidence, there were demonstrable and significant health and environmental benefits to organic agriculture, there remains the question regarding the potential for organic agriculture to meet the global food needs today and in the future.

There is an element of nostalgia associated with organic agriculture, a desire to return to a simpler life, a life more connected to the land. Without doubt, most people in rich countries live in cities and have become disconnected from the sources of their food. Moreover, urbanisation has led to large swathes of countryside – including prime farmland – being developed for

[1] Greenpeace has its strongest base of support in Europe and until recently received almost unconditionally supportive reporting in most European media. As supportive scientific evidence about the safety and environmental benefits of GM crops and food has mounted, and given that more than 15 years of experience with it in North America has not led to the horrors Greenpeace portrayed, the tide seems to be turning. Perhaps more importantly, given the kinds of factors that turn public opinion, Greenpeace's involvement in vandalism of GM crops – such as the July 1999 vandalism in Britain led by Lord Peter Melchett, then the executive director of Greenpeace – has caused a backlash against the group. What especially galvanised media opposition was the vandalism of scientific study field trials, since they were designed to address precisely the kinds of concerns that Greenpeace has always claimed need to be addressed.

housing and commerce – including roads. Urbanisation and people's insatiable appetite for energy, unquestionably, have led to environmental degradation: clearing of forests, draining of marshes, diversion of watercourses, damning rivers, pollution of water and air, and so on. The list is long and familiar to most people. I have touched on this in Section 6.1 and will focus on it again in this chapter. Hence, the 'simpler life, more connected to the land' is attractive. As we shall see, even minimal reflection reveals the much less appealing character of this nostalgic picture.

In these introductory remarks, I have not masked my scepticism about the tenability of the claims made about organic agriculture. This scepticism, however, should not be construed as opposition. I support **an** organic alternative;[2] my scepticism is about the inflated rhetoric, which goes well beyond the evidence and sidetracks progress on the urgent need to reduce the negative impact of agriculture while ensuring food security. Michael Spector (2009) has skilfully demonstrated the negative impacts of what he calls 'denialism' and how the organic agriculture movement engages in harmful denialism. I agree with much of his analysis but take a somewhat softer approach and provide a much deeper examination.

The evidence, as set out so far in this book, and added to below, suggests that organic cannot be **the** only, or even the predominant, alternative to the *status quo*. There is, clearly, potential for increasing the portion of agriculture that is organic and there will be some environmental benefits from doing so, but the overall assessment from the arguments and evidence presented in this book is that non-organic agriculture will continue to be dominant. That means policymakers, economists, environmentalists and citizens will have to identify the least environmentally harmful non-organic agricultural practices; and they will need to identify the practices that will result in the healthiest food possible. This, I argue below, is where the opposition to GM

[2] I live in a rural area and buy my eggs from a neighbour whose hens have the full run of a yard and are few in number; her farm is organic. I have a third of an acre (0.13 hectare) vegetable garden, which is effectively organic but not certified as such since I do not sell my produce. I freeze, jar, dry and root cellar as much of my garden produce as I can. I am under no illusions, however, that my choices in this regard are based on nostalgia and romanticism and are not rooted in empirical evidence regarding environmental or health benefits versus harms. There is a lot to be said for including aesthetic dimensions in one's decision-making but they should not be confused with empirical evidence (see Castle, 2003).

agriculture by advocates of organic agriculture is an impediment to meeting the critical agricultural challenges of the future.

7.1 The environment: conventional, organic and GM agriculture

There is indisputable evidence that conventional farming has a large negative impact on the environment, an impact that is unsustainable. Although recent work (Bahlai *et al.*, 2010) suggests that the environmental impacts of organic vs. synthetic may be less than touted, organic agriculture does appear to have a somewhat lower negative impact, but far from zero. In assessing its negative impact, it is important to separate animal from non-animal agriculture. Non-animal organic agriculture has a much lower environmental impact than conventional non-animal agriculture, and it is here that the greatest mitigation can be achieved by increasing the amount of organic agriculture. Animal agriculture, on the other hand – organic or conventional – is environmentally problematic.

Animal agriculture has a long history. None of the domesticated food animals today could survive without human tending, and it is human tending on the scale of modern agriculture that creates environmental problems that are difficult to mitigate (see Steinfeld *et al.*, 2006). However, one should not lose sight of the fact that even the romantic 'small-scale' organic farm creates environmental problems that are difficult to mitigate, and, it is also worth noting, 'small-scale' organic is continually being replaced by 'large-scale' organic as organic agriculture itself becomes 'big business'. Marketplace success and government subsidies continue to draw in entrepreneurs who understand profits well. There are non-domesticated sources of meat (wild moose, buffalo and deer, for example), which do not require agricultural tending but substituting that source for domesticated sources on the scale required to meet current demand will lead to decimation of the herds, which is just a different negative environmental impact. In effect, one environmental disaster is being replaced by another.

The major environmental impacts of animal agriculture result from characteristics of the animals. Their manure is high in nitrogen, phosphorus, potassium and other plant nutrients such as calcium, magnesium and sulphur. This makes it an excellent fertiliser but also an environmental calamity. Consider just the nitrogen content. Some of the nitrogen is slow-release but most is rapid-release. Much of the rapid-release quickly enters the environment through ammonia volatilisation – conversion to urea, which gases off into the atmosphere as a greenhouse gas. Incorporating the manure into the soil quickly and under cold moist conditions reduces this process but provides a higher level of nitrogen in the soil that ultimately will be converted to nitrate, which enters the groundwater. This pollutes the groundwater but eventually that groundwater reaches rivers and lakes, ammonia volatilisation occurs, and the nitrogen gases into the atmosphere. Moreover, animals' urine is high in urea; their flatulence is high in methane, which is released into the atmosphere (one of the largest sources of atmospheric methane); their feed requirements from plants (or from other animals, which ultimately derives from plants) are high; their water requirements are high; and on the list goes. Organic agriculture can do little about these characteristics. It might reduce, or eliminate, the use of antibiotics and hormones. It might feed the animals organically produced plant materials. It might allow them to graze in pastures, reducing feedlot demand. Nonetheless, ultimately, barnyard animals will produce, per animal, the same amount of faecal material, urine and flatulence, and will require the same amount of food and water.

The solution to detrimental environmental effects of animal agriculture is not going to be found by embracing organic agriculture. As already suggested, an obvious solution is a reduction of meat consumption (and eggs and dairy products) but there are no grounds for believing that will occur on the scale required. Indeed, as more of the world's poor people become even slightly more affluent, demand for animal products will increase. Consequently, placing one's hope for mitigation of agriculturally caused environmental degradation on a pro-vegetarian diet is a risky strategy, as risky, in fact, as placing one's hope on a pro-abstinence and anti-condom strategy for combating HIV/AIDS. Where the latter has been tried, success has been limited, at best, and failure more common; there is little reason to expect a pro-vegetarian strategy to be any more successful. Both sexual propensities and a passion for meat are primordial; pro-vegetarian or pro-abstinence strategies court failure by underestimating this factor. I am a vegetarian (or, more accurately, a piscatarian since I eat fish and other seafood but no land animals). I *think* it is a healthier diet (though I accept that the evidence supporting this is sketchy at best and changes frequently). With greater certainty, I think that, if vegetarianism were widespread, it would enhance environmental sustainability and would feed more people adequately, using less land (and, secondarily, animal suffering would be reduced). Nonetheless, I have no illusions about vegetarianism becoming widespread or about it being more than a very small part of mitigating the environmental impact of animal agriculture.

The heart of this issue, manifestly, centres on quantity. A few cows, pigs and the like on a 50-acre (20-hectare) property will have minimal environmental impact but will also not make a meaningful contribution to meeting current meat demand. Increase the numbers (to, say, 100 cows and 30 pigs) and the environmental impact increases. If that were the only animal agriculture in the region, the impact would not be significant; that, however, is seldom the case. From an environmental perspective, farm size is irrelevant; a large number of 50-acre farms with 100 cows and 30 pigs in a region is indistinguishable, in terms of the environmental impact, from a 1,000-acre farm with 2,000 cows and 600 pigs. The attraction of organic agriculture comes from an image of the old-time family farm with a few cows, pigs and goats roaming free and a few chickens pecking at bugs in the barnyard and scavenging (recycling) farmhouse food wastes. In this picture, the manure is used to fertilise the vegetable garden and cropland. In the early days of organic agriculture, this image may have conformed to reality; increasingly, the sources of organic products are large-scale operations and, indeed, to meet the demand for organic animal products, organic agribusinesses are arising. Without a reduction in demand, the overall numbers of agricultural animals will be constant – in fact, will grow – whatever the specific mix of conventional and organic animal agriculture. Moreover, the use of manure for fertiliser – a practice that in no way is restricted to organic farms – does not change the environmental impact on groundwater pollution. Crops use a very small portion of the available nitrogen, phosphorous and potassium; the rest finds its way into the atmosphere or groundwater, and, in the case of nitrogen, from groundwater into rivers and lakes and into the atmosphere as greenhouse gases.

The upshot of all this is that organic animal agriculture will at best make a minimal contribution to mitigating agriculturally caused environmental degradation; the underlying factors are consumer demand and animal physiology, neither of which is addressed – probably cannot be addressed – by embracing organic agriculture. This is a rather serious fault line in the idealised conception of the benefits of organic agriculture. There are other serious fault lines as well.

Another of the nostalgic attractions of organic agriculture is the mystique around using one's own seeds from year to year and the use of 'heritage' seeds. It seems more natural to allow the plants from one year to provide the germs of life for the next, rather than depend on large, for-profit, commercial, seed companies. Moreover, the term 'heritage seed' harkens back to the way things were 'naturally' before contamination by human activity. Neither holds up to even minimal scrutiny in organic agriculture.

The need to be clear about the scope of 'heritage seed' has already been discussed in Section 6.2. Although there are demonstrable reasons for preserving heritage seed stocks, 'heritage' is not equivalent to 'natural' (not manipulated by humans). On the issue of retaining seeds, organic (or conventional) farmers who wish to retain seeds from year to year are restricted to open-pollination species and varieties. Hybrids will not breed true in the next generation. Recall from the population genetics background that during meiosis (gamete formation) pollen and ovules with only one of *A* or *a* will be formed. The ratio of *A* pollen and *A* ovules to *a* pollen and *a* ovules is close to 1:1 (0.5:0.5 in population genetic notation). If we assume close to random pollination, the segregated *A* and *a* alleles will recombine in the fertilised ovules as follows:

	A	*a*
A	*AA*	*Aa*
a	*Aa*	*aa*

Hence, a field of hybrids will produce 50 per cent non-hybrid seed. A farmer will not know by inspection which are hybrid seeds, and any laboratory procedure would have to examine each of the seeds to sort them into *AA*, *Aa* and *aa* – a procedure that would be expensive and time-consuming. That is why farmers who want to grow a specific hybrid plant because it manifests a desired trait or exhibits hybrid vigour (heterosis) will buy seed yearly from a seed company. Almost all commercial corn (maize) exhibits heterosis, which gives it one of its trait values.

Hybridisation is exceptionally important in plant agriculture, so let's look a little more closely at it. Heterosis is obviously an advantage, whereas loss of viability in F2 (outbreeding depression: lower fitness in F2 than either parental stock) is not. But that is what a farmer who retains hybrid seed faces. A cursory examination of seed catalogues, even those that cater for organic farmers, reveals the large number of hybrid seed varieties available, and organic farmers, like conventional farmers, take full advantage of the beneficial traits of these hybrids. Many farmers, organic and conventional, retain seed from year to year, but most also use hybrids and buy guaranteed seed every year. This mix is to be expected and in no way diminishes a farmer's claim to be organic, but it does undermine the nostalgic view of the simple-life farmer who is independent of the commercial seed world.

7.2 Health: evidential lacunae

The health benefits of organic agriculture are presumed to result from no exposure to synthetic pesticides, herbicides, fertilisers, hormones and the like (the suggestion being that synthetic chemicals are more detrimental to one's health than natural chemicals – or, at least, the natural ones used by organic farmers such as rotenone and pyrethrum dusts and *Bt* sprays). Current evidence, however, does not support the claimed health superiority of organically grown food to non-organically grown food, whether non-GM or GM. This is not surprising because contrary to the simplistic claims made by advocates of organic agriculture, empirical research in this domain is exceptionally challenging. As Magkos *et al.* (2006) note in passing (see also Gilbert-López *et al.*, 2009),

> Addressing food safety of organic versus conventional produce is difficult, especially in the face of limited and conflicting data. In order to carry out a valid comparison between organic and conventional food products, it is required that the plants be cultivated in similar soils, under similar climatic conditions, be sampled at the same time and pre-treated similarly, and analyzed by accredited laboratories employing validated methods (Kumpulainen, 2001). In terms of foods of animal origin, animals would have to be fed on plants meeting the above production criteria. (p. 45)

Health claims typically focus on nutrition, health enhancements and health detriments: that is, on the degree to which a food delivers nutrients essential for sustenance (e.g. vitamins, minerals, amino acids), enhances health (e.g. high fibre content, antioxidants, omega-3 fatty acids), or diminishes health (e.g. mercury levels, carcinogens, neuro-toxins). Quantifying the nutrient content of foods is significantly easier than determining the enhancement or diminishing of health. The challenges with respect to enhancement and diminution are fivefold. First, there are numerous interacting factors involved. The health status and changes to health status depend on an individual's genetics, lifestyle (the impact, for example, of high fat consumption will be different for an individual who leads a sedentary life than for a physically active individual), climate, exposure and immune response to pathogens, composition of diet (a yearly diet high in red meat compared to a largely vegetarian diet will result in different biodynamical systems), non-dietary exposure to chemicals, and the list goes on. The sheer number of factors makes causal attribution (i.e. a claim that consumption of *x* level of *y* chemical over *t* time will cause *z* health effect) or prediction dubious, even if qualified with a probability measure. That all these factors interact renders almost any claimed cause-and-effect relationship unreliable. Since each individual will have a different combination of this large array of factors, the effect of a change in some aspect of one factor (dietary pesticide ingestions, for example) will be idiosyncratic (specific to that person). Randomised controlled trials (RCTs) are, in principle, supposed to tame this heterogeneity by comparing two groups that contain the same heterogeneous mix of individuals. The only difference, it is assumed, is the element under study. Hence, any difference in outcome must be caused by the element under study. There are mathematical and conceptual reasons to be sceptical that RCTs provide reliable causal knowledge (see Howson and Urbach, 1989; Kravitz *et al.*, 2004; Salsburg, 1993; Thompson, 2010, 2011a, 2011b; Upshur, 2005; Worrall, 2002). The numerous contradictory RCTs that have appeared in the last three decades, and the numerous changes in medical advice (including dietary advice) based on RCTs, signal that research in this domain is difficult, that generation of stable knowledge is illusive, and that RCTs are not a robust research method, even though, in many cases, they may be the only applicable method.

Second, the available techniques for detecting the presence of relevant chemicals (natural and synthetic, and harmful and beneficial) often lack the required precision. Third, there is a difference between the presence, or use, of chemicals during production and the presence of those chemicals in the food as consumed. Fourth, research on the health effects of relevant chemicals is impoverished and is often ambiguous and/or contradictory. Fifth, different processing methods result in changed and different chemical profiles (the ubiquitous process of heating food can cause some chemicals to gas off and others to be transformed), yielding different health impacts. Changes to the chemical profile during processing (commercial or consumer) can be beneficial, harmful or neutral to health. Moreover, the same food in the same end-consumer kitchen will be prepared differently on different occasions, resulting in different chemical profiles and, hence, different health impacts – including the nutritional. This alone renders research on health impacts challenging and expensive. A steak from an organically raised and fed heifer when grilled may contain more known carcinogens than a stewed steak from a conventionally raised and fed heifer. Grilling or roasting involves very high surface temperatures (250 °C or higher). At these temperatures, a non-enzymatic reaction occurs; sugars and amino acids interact in a Maillard reaction, which begins to occur at around 160 °C and above. This causes the observed browning of the surface and the change in flavour. The same reaction occurs with all grilled meats (beef, fish, pork and chicken) and many grilled vegetables, especially when brushed with oil. Many of the compounds created through this process (aromatic hydrocarbons and heterocyclic amines, for example) are known carcinogens. Research to date does not suggest that one should remove grilled meats from the diet; too little is known about exposure thresholds and compensatory physiological mechanisms. What it does suggest is that simplistic comparisons of the health impacts of organically and conventionally farmed food are entirely unreliable.

In light of this, anyone who claims that organic food has been shown to be more nutritious, less harmful or more health enhancing than conventionally produced food is being disingenuous – or perhaps has been duped. Also in light of the complexity of the task, it is not surprising that careful reviews of the research literature consistency fail to find evidence of a nutritional or health difference between organically and conventionally produced foods.

Faidon Magkos *et al.* (2006), in a recent, exceptionally comprehensive and balanced review, reviewed upwards of 400 articles on organic food and health. The abstract reads:

> Consumer concern over the quality and safety of conventional food has intensified in recent years, and primarily drives the increasing demand for organically grown food, which is perceived as healthier and safer. Relevant scientific evidence, however, is scarce, while anecdotal reports abound. Although there is an urgent need for information related to health benefits and/or hazards of food products of both origins, generalized conclusions remain tentative in the absence of adequate comparative data. Organic fruits and vegetables can be expected to contain fewer agrochemical residues than conventionally grown alternatives; yet, the significance of this difference is questionable, inasmuch as actual levels of contamination in both types of

> food are generally well below acceptable limits. Also, some leafy, root, and tuber organic vegetables appear to have lower nitrate content compared with conventional ones, but whether or not dietary nitrate indeed constitutes a threat to human health is a matter of debate. On the other hand, no differences can be identified for environmental contaminants (e.g. cadmium and other heavy metals), which are likely to be present in food from both origins. With respect to other food hazards, such as endogenous plant toxins, biological pesticides and pathogenic microorganisms, available evidence is extremely limited preventing generalized statements. Also, results for mycotoxin contamination in cereal crops are variable and inconclusive; hence, no clear picture emerges. It is difficult, therefore, to weigh the risks, but what should be made clear is that 'organic' does not automatically equal 'safe.' Additional studies in this area of research are warranted. At our present state of knowledge, other factors rather than safety aspects seem to speak in favor of organic food. (p. 23)

They conclude:

> The asserted health benefits are impossible to quantify and do not seem, as yet, to compensate for the increased price. It is also important to note that, at present, there is no scientifically tenable evidence that any differences observed between organic and conventional food would lead to any objectively measurable effects on human health. In fact, health benefits resulting from the consumption of a specific food or food ingredient are not unanimous, but most probably depend on the genetic background, dietary habits, and overall lifestyle of an individual (Eckhardt, 2001). Further, there are many limitations and lots of uncertainty in assuming that increasing the dietary level of any compound will necessarily improve health (Trewavas and Stewart, 2003). It has been suggested, therefore, that individual metabolism, as it relates to health, predisposes to disease, or other health outcomes, should guide future agriculture toward foods for improved health and nutrition (Watkins *et al.*, 2001).
>
> There is currently no evidence to support or refute claims that organic food is safer and thus, healthier, than conventional food, or vice versa. Assertions of such kind (Colborn *et al.*, 1996; Avery, 1998; Rogers, 2002) are inappropriate and not justified, and remain groundless not only due to ethical considerations but also because of limited scientific data. The selective and partial presentation of evidence serves no useful purpose and does not promote public health. Rather, it raises fears about unsafe food. (p. 47)

The lack of empirical evidence supporting claims that organic food is healthier (or, for that matter, safer) is not something recently uncovered. The House of Commons Agricultural Committee (UK) reported in 2001:

> This is not to accuse the organic movement of misleading the public but it is perhaps true that the public has a perception of organic farming that is, at least partly, mythical. **We believe it important that the claims can be tested and verified in order that consumers know what they are really buying**. The statement from the Food Standards Agency (FSA) in August 2000 that it "considers that there is not enough information available at present to be able to say that organic foods are significantly different in terms of their safety and nutritional content to those produced by conventional farming" raised a furore, but illustrates the limits of claims which can be scientifically sustained. Research to sustain or quantify the claimed benefits of organic farming is badly needed.

Similarly, the Parliamentary Information and Research Service of Canada (Forge, 2004) reported: 'Although beneficial to the environment, organic farming methods are not guaranteed to produce healthier foods than those produced by conventional farming methods . . . The label "organic" does not provide any guarantee of a product's quality and nutritional value.'

Conspiracy theories about government motivations are numerous, so government reports such as these are often viewed with scepticism, a stance, however, that cannot be taken with the Magkos *et al.* review. Also, it does not apply to the FSA, which the House of Commons committee cites; the FSA is an independent government department set up by an Act of Parliament in 2000 to protect the public's health and consumer interests in relation to food. Its current posting on organic food (www.food.gov.uk/foodindustry/farmingfood/organicfood/#h) explicitly states: 'The Agency is neither for nor against organic food. Our interest is in providing accurate information to support consumer choice.' It is also explicit that, 'The available evidence shows that the nutrient levels and the degree of variation are similar in food produced by both organic and conventional agriculture.'

There are also compelling scholarly reviews prior to the Magkos *et al.* review of 2006. For example, Bourn and Prescott in a 2002 article wrote:

> Given the significant increase in consumer interest in organic food products, there is a need to determine to what extent there is a scientific basis for claims made for organic produce. Studies comparing foods derived from organic and

> conventional growing systems were assessed for three key areas: nutritional value, sensory quality, and food safety. It is evident from this assessment that there are few well-controlled studies that are capable of making a valid comparison. With the possible exception of nitrate content, there is no strong evidence that organic and conventional foods differ in concentrations of various nutrients. Considerations of the impact of organic growing systems on nutrient bioavailability and nonnutrient components have received little attention and are important directions for future research. While there are reports indicating that organic and conventional fruits and vegetables may differ on a variety of sensory qualities, the findings are inconsistent. In future studies, the possibility that typical organic distribution or harvesting systems may deliver products differing in freshness or maturity should be evaluated. There is no evidence that organic foods may be more susceptible to microbiological contamination than conventional foods. While it is likely that organically grown foods are lower in pesticide residues, there has been very little documentation of residue levels. (p. 1)

The above quoted reports and reviews seriously undermine any confidence in claims that organic food is healthier, more nutritious or safer than conventional food. The current state of evidence is something of a stalemate. Given that organic food is more labour intensive, that there are yield concerns (more on this later), and that consumers have to pay more for it, the onus of proof that there are benefits that offset these factors, in my view, falls to the organic food producers. Organic food has been promoted in the last several decades by NGOs, by the organic growers' organisations, and through government subsidies and policies. The promotions invariably cite health and safety benefits over conventionally produced food, especially with respect to carcinogens resulting from pesticide and herbicide use. However, in an article in the *Proceedings of the National Academy of Sciences of the USA*, Ames *et al.* (1990) have pointed out that natural carcinogens (carcinogens that plants produce) swamp the minute traces from pesticides and herbicides:

> The U.S. Food and Drug Administration (FDA) has assayed food for 200 chemicals including the synthetic pesticide residues thought to be of greatest importance and the residues of some industrial chemicals such as polychlorinated biphenyls (PCBs) [they cite Gunderson (1988)]. The FDA found residues for 105 of these chemicals: the U.S. intake of the sum of these 105 chemicals averages about 0.09 mg per person per day, which we compare to

> 1.5 g of natural pesticides (i.e., 99.99% natural). Other analyses of synthetic pesticide residues are similar [they cite Nigg *et al.* (1990)]. About half (0.04 mg) of this daily intake of synthetic pesticides is composed of four chemicals [they cite Gunderson (1988)] that were not carcinogenic in rodent tests: ethylhexyl diphenyl phosphate, chlorpropham, malathion, and dicloran [they cite Gold *et al.* (1984) and Treon *et al.* (1953)]. Thus, the intake of rodent carcinogens [compounds that at high doses have produced cancers in rodents] from synthetic residues is only about 0.05 mg a day (averaging about 0.06 ppm in plant food) even if one assumes that all the other residues are carcinogenic in rodents (which is unlikely). (pp. 7779–7780)

Hence, even if we could eliminate all non-naturally occurring chemicals from food through organic farming (and the evidence suggests that is unlikely), the reduction in potentially harmful chemicals in our food will be miniscule (0.05 mg per person per day of non-natural potentially harmful chemicals compared to 1,500 mg of the same mix of chemicals occurring naturally – so, of our average daily intake of these chemicals, 1,499.95 mg (out of the 1,500 mg: i.e. 99.9967 per cent) naturally occurs in plants we consume (see also Céline Menard *et al.*, 2008). Moreover, as commented on above:

> The cooking of food is also a major dietary source of potential rodent carcinogens [chemicals that, at exceptionally elevated levels, have been found carcinogenic in rodents]. Cooking produces about 2 g (per person per day) of mostly untested burnt material that contains many rodent carcinogens – e.g., polycyclic hydrocarbons [they cite Clarke and Macrae (1988) and Furihata and Matsushima (1986)], heterocyclic amines [they cite Sugimura (1988) and Takayama *et al.* (1987)], furfural [they cite Maarse and Visscher (1989) and Stofberg and Grundschober (1987)], nitrosamines and nitroaromatics [they cite Ames *et al.* (1987 and 1990) and Beije and Möller (1988)] – as well as a plethora of mutagens [they cite Furihata and Matsushima (1986), and a special issue of *Environmental Health Perspectives* (1986)]. Thus, the number and amounts of carcinogenic (or total) synthetic pesticide residues appear to be minimal compared to the background of naturally occurring chemicals in the diet. Roasted coffee, for example, is known to contain 826 volatile chemicals [they cite Maarse and Visscher (1989)]; 21 have been tested chronically and 16 are rodent carcinogens [they cite Gold *et al.* (1984, 1986 and 1987)]; caffeic acid, a nonvolatile rodent carcinogen, is also present. A typical cup of coffee contains at least 10 mg (40 ppm) of rodent carcinogens (mostly caffeic acid, catechol, furfural, hydroquinone and hydrogen peroxide). The evidence on coffee and

> human health has been recently reviewed, and the evidence to date is insufficient to show that coffee is a risk factor for cancer in humans [they cite Clarke and Macrae (1988), and the national Research Council (1989)]. The same caution about the implications for humans of rodent carcinogens in the diet that were discussed above for nature's pesticides apply to coffee and the products of cooked food. (Ames *et al.*, 1990, p. 7780)

Given this, the case for a meaningful health benefit from organic food is highly suspect. No doubt some individuals will opt for the miniscule reduction that organic food *might* deliver at source (0.0033 per cent) and, in an open and pluralistic society, that choice should be available. What is essential is that consumers be made aware that the difference at source between organic and conventional is minuscule and that the effects of cooking food will further diminish the difference (the sautéed organic onion and conventional onion, for example, will differ imperceptibly in health compromising chemicals). Regrettably, that is not the information consumers are given.

7.3 The problem of yields

There has been considerable debate in the past decade about yields from organic agriculture and whether organic farming could, on the current agricultural footprint, feed the world. Catherine Badgley (2007) and her colleagues, for example, claim:

> Our models demonstrate that organic agriculture can contribute substantially to a more sustainable system of food production. They suggest not only that organic agriculture, properly intensified, could produce much of the world's food, but also that developing countries could increase their food security with organic agriculture. The results are not, however, intended as forecasts of instantaneous local or global production after conversion to organic methods. Neither do we claim that yields by organic methods are routinely higher than yields from green-revolution methods. (p. 94)

Goulding and Trewaves (2009) disagree:

> We have examined the literature basis of these claims particularly on wheat. There are many omitted references that indicate organic yields are substantially lower than Badgeley *et al.* (2007) indicated. There are calculation errors in some of the references used by Badgeley *et al.* (2007). Also Badgeley *et al.* (2007) are equating organic procedures only with the use of either

manure or cover crops and are ignoring certified organic procedures that prohibit synthetic pesticide use. We have also examined the claims by these authors that there is sufficient N fixed to provide for fertiliser and have found that mineralisation levels are wrongly equated with the N appearing in seed yield. We agree with Badgely *et al.* (2007) that maintenance of organic material in soil is important but consider that this is not a specific organic procedure. There would be insufficient food for the world population provided by global organic farming.

Moreover, the United Nations and various government sources do not support the position of Badgley and her colleagues.

This, like assessments of health effects, is difficult terrain. Recall the comments of Magkos *et al.* (2006) cited in Section 7.2:

> Addressing food safety of organic versus conventional produce is difficult, especially in the face of limited and conflicting data. In order to carry out a valid comparison between organic and conventional food products, it is required that the plants be cultivated in similar soils, under similar climatic conditions, be sampled at the same time and pre-treated similarly, and analyzed by accredited laboratories employing validated methods (Kumpulainen, 2001). In terms of foods of animal origin, animals would have to be fed on plants meeting the above production criteria. (p. 45).

Although made with respect to health claims, these conditions apply equally to yields; determining and comparing yields is exceptionally challenging. Factors such as the specific crop, the regional soil and climate profile, and transitional costs bedevil any simple comparison. One cannot compare the yield achieved for tomatoes grown in an area with nutrient-rich soil, ideal rainfall patterns and days of sun with the yield for the same tomato variety grown in an area with marginal soil, unpredictable rainfall and fewer sun days. A difference in any one of these factors is sufficient to undermine the comparison. And the reality is that a large amount of food is grown under marginal conditions. Organic agriculture will inevitably experience lower yields on these soils than conventional agriculture. GM, as previous chapters have demonstrated, offers a way to reduce – in some cases eliminate – the use of pesticides and herbicides, to maintain yield in drought-ridden areas and to reduce a plant's nutrient requirement, such as nitrogen. Notwithstanding these complexities, the evidence that has been generated does seems clear:

organic yields are lower, sometimes dramatically lower, than conventional yields.

In a pluralistic and open society, there is clearly a place for organic farming in the overall solution to the environmental concerns, but it will remain a boutique part of agriculture. Many EU countries have promoted organic farming and provided significant subsidies; nonetheless, the organic sector has remained very small. Acres of organic farming to total acres of farming vary from a high of 11.1 per cent in Austria (US$9.9 million in subsidies; US$27.5 per acre) to 0.4 per cent in the USA (Denmark is 5.5 per cent, Canada 0.58 per cent, France 1.9 per cent, Italy 7.7 per cent, Switzerland 7.4 per cent, and Sweden 6.4 per cent); data on total acres in agriculture are from the Food and Agriculture Organisation of the United Nations, and acres in organic farming is from the International Federation of Organic Agricultural Movements. There are clear demand-for-food and economic reasons for these very small percentages.

7.4 GM and organic: the false dichotomy

In agriculture, conventional and organic are principally terms describing methods of farming. Specifically, organic farming avoids many of the inputs used in conventional farming, such as most pesticides, herbicides and synthetic fertilisers. Pesticides that are used, such as rotenone, pyrethrum and *Bt*, and fertilisers, such as cattle manure, are deemed 'natural', in contrast to 'synthetic' pesticides and fertilisers, such as 2,4-D and nitrogen derived from fossil fuels. Regulations and certification standards for use of the term 'organic' differ by country but they all recognise this basic difference.

As we have seen, organic farming methods are, in part, promoted as a way to address concerns about the environmental impact of conventional farming and concerns about the health impacts of synthetic chemical residues on or in food. These reasons for promoting organic farming do not, in any straightforward way, entail that organic crops are better than GM crops. That is because organic crops and GM crops are types of plants, not methods of farming. One could use organic methods or conventional methods in growing GM crops.

There is, therefore, a two-by-two comparison occurring, as shown in the following table:

	Organic farming	Conventional farming
Chemicals used	• Naturally occurring chemicals only	• Synthetic • Naturally occurring chemicals
Seed/plant type used	• Open pollinated • Hybrids	• Open pollinated (less used) • Hybrids • GM

Note that hybrids are permitted and used in organic farming (and there is no requirement to label food as hybrid plant in origin). Also, the hybrids used are the result of human manipulations of nature; this involves the application of population genetics, not molecular genetics. The hybrid crops in widespread use in both conventional and organic farming are not products of nature but of humans. So what difference could there be between GM human-created crops and hybrid human-created crops? The standard answer appeals to naturalness. Hybrid crops are natural whereas GM crops are not. Obviously, 'natural' here cannot mean arose naturally, since they are the direct result of human intervention in nature. 'Natural' appears to mean that fertilisation and development occur in accordance with the laws of nature. Human intervention is limited to controlling what pollen is associated with what ovaries. There is no alteration of the laws of nature. This, you will recall, was the claim made in the Supreme Court of Canada judgement in the *Pioneer Hi-Bred Ltd.* v. *Canada* case. Since, as a matter of historical fact, the hybridisation did not arise naturally, the underlying assumption must be that it could have arisen naturally; nothing a seed company does goes beyond what could have occurred in the absence of human intervention. The probability, however, that nature would have generated, over time, the specific hybrids is very low. Of course, it is not zero; there is always a low probability that the original plants involved in the hybridisation may become sufficiently geographically close for cross-pollination to occur, that effective pollination occurs (by birds, bees and wind, for example), that self-pollination is naturally inhibited (mimicking human detasselling of corn, for example), that the resulting plants can compete as least as successfully, without human tending, against weeds and pests as the parental non-hybrid plants, and so on. So, the claim seems to be that human-generated hybrid plants are 'natural' because natural processes 'could have produced them' no matter how unlikely.

Acceptance of the possibility but improbability of non-human-generated hybrid crops identical to those widely planted on conventional farms and organic farms renders the case against GM as unnatural less compelling. Although improbable, natural processes could have produced *Bt* corn; mutations, substitutions of nucleotide sequences and bacterial alteration of a plant's DNA occur all the time. Indeed, the technique used to get the DNA sequence that codes for the production of the *Bt* endotoxin into the DNA of the crops exploits the fact that *Agrobacterium tumefaciens, in nature*, alters the DNA of plants – causing, in this case, crown gall. Nature is constantly producing novel forms of life, constantly giving rise to new genes and constantly engaging in, metaphorically speaking, trial-and-error experiments.

Van Valen (1973) named one aspect of the dynamics of evolution the Red Queen hypothesis (frequently now called the Red Queen phenomenon or effect). It is an allusion to the Red Queen in Lewis Carroll's *Through the Looking-Glass*. Alice, who has just finished running very fast with the Red Queen only to find herself under the same tree as when they started, expresses surprise:

> Alice looked round to her great surprise. "Why, I do believe we've been under this tree the whole time! Everything's just as it was!"
>
> "Of course it is," said the Queen. "What would you have it be?"
>
> "Well, in *our* country," said Alice, still panting a little, "you'd generally get somewhere else – if you ran very fast for a long time, as we've been doing."
>
> "A slow sort of country!" said the Queen. "Now, *here*, you see, it takes all the running *you* can do, to keep in the same place. If you want to get somewhere else, you must run at least twice as fast as that!"

Others have used the metaphor of an evolutionary arms race to capture this dynamic. A predator and prey co-evolution is an illustration. The prey acquires (perhaps through a mutation) a way of eluding a predator or expressing a toxin to kill the predator. The predator then acquires a new way of detecting the prey or becoming tolerant to the toxin. The prey evolves yet another trait and again the predator evolves a response, and on it goes. From bacteria to complex multicellular organisms, from cells to plants and animals, living things constantly change; the constant production of variation is essential for evolution and evolution is essential for long-term survival of a species lineage.

Many changes are detrimental to an organism's survival and will not be replicated in subsequent generations. Occasionally, a variation provides an advantage – more efficient use of nutrients, increased protection from predators, more success in detecting and/or capturing prey, for example – and is replicated in subsequent generations. Selection acting on heritable variation is the main mechanism that allows organisms to remain fit (survive) in a constantly changing environment of other organisms and physical conditions. They must constantly change to stay fit. A critic of GM might claim that these changes are 'natural'. They will occur – and, of course, did for millions of years – without humans. GM plants are not 'natural' but human designed. Although it is undeniably true that they are human designed, it is irrelevant. If human-generated hybrid plants are 'natural' because natural processes 'could have produced them' no matter how unlikely, then so are GM plants. Consequently, the difference between GM and non-GM agricultural plants cannot be human manipulation that supersedes nature; even most open-pollinated agricultural plants have undergone intensive human selection and manipulation.

A slightly different tack focuses on the fact that domestication through selection does not involve *designing the traits*; GM does. One needs to be clear about what is being packed into the expression, 'designing traits'. 'Design', of course, normally has a broad scope and would include a seed company designing wheat that has traits making it more suited to the climate of the plains of central Canada. This may involve searching for a mutant plant or hybridising plant varieties that would not normally interfertilise. Many organic farmers grow such selected mutants and hybrid plants, and a substantial amount of organic food is derived from them. So 'design' in this context must be more restricted than its ordinary language use in order to effectively demarcate GM from non-GM; in effect, its meaning must be narrowed to 'molecular designing'. But then, 'designing traits' is merely a substitute (a synonym) for 'molecular manipulation of traits'. Since the two expressions are equivalent by definition, nothing is gained by using one rather than the other; to say that designing traits is unacceptable is merely to use different words to say molecular manipulation of traits is unacceptable. That, however, means that no criterion of demarcation has been given either.

So, it is clear that what is offensive about GM plants to the organic movement centres entirely on the fact that it is a molecular genetic manipulation; it has nothing to do with human intervention in nature, nothing to do with imposing human desires on the type and number of a particular plant or animal variety (designing plants and animals), nothing to do with being 'natural', nothing to do with eschewing science and technology, nothing to do with the economics of agriculture, and so on. In essence, the organic position can be distilled to a simple contrast. If nature produces a mutant with a trait that enhances the plant from the human perspective, selecting for that trait is OK – even if that mutation would not have had a chance of survival without human intervention and even if the human intervention results in plants with that trait swamping those without it. If, however, humans directly alter the DNA to cause the same mutation, it is not OK. At this point, the aversion to GM is obviously ideological. In Section 4.2 the core ideological objection to GM was examined and deficiencies in the assumptions and arguments on which it rests were identified.

In light of the examination, in previous chapters, of the ideological and pragmatic issues arising from agricultural biotechnology and the congruence of the goals of organic agriculture and those of GM agriculture, continued opposition of organic to GM (and vice versa) seems unjustifiable. Continued opposition to GM by advocates of organic agriculture (or vice versa), in the face of the fact that both strive to achieve the same goals, would suggest that more is motivating the opposition than concerns about the environment, health, manipulating life and the like. One can reasonably speculate that protection of market share is at least a meaningful part of the motivation to continued opposition. This is the nature of commerce, marketing, developing market share, and all the other aspects of regulated free-market economics. There is nothing wrong with this; what is wrong is to pretend that the motivation is not commercial but moral.

8 Impacts on low- and middle-income countries

Poverty, farming and colonial legacies

This chapter moves the focus away from rich countries to low- and middle-income countries, in particular low-income sub-Saharan African countries. Having been many times to rural areas of western Kenya, I have a personal interest in the GM debate in Africa and in the larger agricultural debates as they relate to farming and poverty in sub-Saharan African countries. I will advance the case that the principal harm arising from GM agriculture in African countries has been, in fact, the unconscionable way in which they have been denied the benefits of scientific and technological advances in agriculture. The same is frequently true of the benefits of science and technology in environmental remediation, in aquaculture and in medicine – especially pharmaceuticals and diagnostic technology such as antiretrovirals for treatment of HIV/AIDS. In the case of medicine, the rich-country outcry has been vociferous, and action to remedy the situation has been forthcoming. Not so with science and technology in agriculture.

In most low- and middle-income countries population growth has outstripped the means of subsistence, and individuals live in various states of poverty, starvation, inadequate nutrition and poor health; the Malthusian dynamic is real. There are no completely reliable data on world poverty. Perhaps the best data are found in the World Bank's 2005 *World Development Indicators*. Globally, in 2001, 2.7 billion people lived on less than US$2 per day. This means 50 per cent of those in low- and middle-income countries are very poor and, as a result, are usually malnourished. Recently, China and India have experienced declines in the proportion of very poor people. Tragically, sub-Saharan Africa has experienced an increase, and that in spite of the billions of dollars in aid that have been poured into African countries during the last decade. The challenge set by the United Nations Millennium Development Goals is to reduce by half the proportion of people who suffer from hunger, measured by malnutrition, by 2015. To achieve this goal means increasing the food resources consumed by 2 billion people, at a minimum. Although the Organisation for Economic Co-operation and Development (OECD) continues to forecast an increase in world agricultural output, the increases fall far short of what will be needed to halve the number of malnourished individuals by 2015. And, halving the number by 2015, although an ambitious goal, leaves half the problem to be solved after that date.

Producing food on that scale risks environmental catastrophe. A theme of this book has been that, as has been the case over the last 200 or so years, science and technology will be a significant element in increasing the supply of food and decreasing the environmental impact of agriculture. This will include continued use of selective breeding and hybridisation but must include molecular genetic technologies.

Regrettably, for the most part, sub-Saharan African countries – except for South Africa – have rejected or at least resisted, the use of GM crops. In light of (1) the benefits described in Chapter 5, (2) the 15 years of widespread use of GM crops and foods in North America – during which time no new risks have emerged – and (3) the critical, urgent need in African countries for increases in the quantity and quality of food, this resistance, to put it mildly, is, at first glance, surprising and troubling. As indicated in previous chapters, although there are challenges (risks) arising from GM agriculture, there is nothing that stands out as different from the challenges faced with every aspect of commerce and, likely, every human activity – from the use of fossil fuels to disposing of human organic waste. Indeed, GM seeds and crops present a more positive profile than most. For these reasons, GM seeds and crops are increasing in use in the Americas and in emerging economies such as China, India and South Africa. **Given all of this, why have sub-Saharan African countries, except South Africa, been so resistant to GM crops? Why are these African countries yet again falling behind?** The depressing fact is that rich nations bear a significant amount of the blame.

Robert Paarlberg in his recent, excellent book (2008), *Starved for Science: How Biotechnology Is Being Kept Out of Africa*, develops a compelling case that these African countries are victims of a rich-world indulgence. To crystallise his point, he compares attitudes in rich countries towards medical biotechnology, on the one hand, with agricultural biotechnology on the other. There is very little protest, in rich countries, against the GM of *E. coli* or goats to produce recombinant insulin and other medical and industrial products, but there is considerable protest directed against GM crops, and foods derived from them.

His explanation of this apparent inconsistency is that in rich countries quality food currently is abundant and cheap; few will die or be malnourished because GM crops and food are resisted.[1] People in rich nations, however, do get sick and do die of disorders and diseases; to reject medical biotechnology will have an immediate and dramatic negative effect on the well-being of those in rich nations.

In short, rich nations can afford to engage in esoteric debates about GM food because, for the present at least, little depends on the outcome; they cannot, do not and will not, engage in debates about GM organisms in medicine because their lives will be significantly negatively affected by a slowing of GMO research and development, and production of recombinant medical products. Moreover, the debates and policy decisions in rich countries – especially European countries – include protectionist goals. Governments of the 27 member counties, the European Parliament (representing the people of Europe), the Council of the European Union (EU) (representing national governments), and the European Commission (representing the common EU interest) have, at various times, used the GM debate as a cover for decisions that have more to do with trade than safety. This could be seen as standard operating practice in the rich-country context, but sub-Saharan African countries are not equal players in that context and their starving populations are collateral damage. If he is correct, and I along with a host of others believe he is, 'hypocrisy' and 'turpitude' are about the only words that are appropriate to describe the current debates about GM crops and the very harmful spillover of those debates to some of the most vulnerable individuals on the planet – the extreme poor in sub-Saharan African countries.

Paarlberg's examination gives rise to another question: why have African countries allowed themselves to be so influenced when China, India, South Africa and South American countries – all emerging economies – have not? Some quickly identifiable factors provide a substantial part of the answer. The legacy of the colonial period, for example, cannot be ignored, nor can the current influence of the NGOs from post-colonial (mainly European) countries that currently work in Africa. Colonial control did not end with independence; an unhealthy and unsustainable dependence on foreign aid continues, and,

[1] It is worth underscoring that, except in the EU, this resistance has not affected the dramatic increase in the planting of GM crops and the production and marketing of foods containing GM plant material, and, as noted, political and policy resistance in Europe is diminishing quickly.

as a result, the economic and social control by rich countries continues. In addition, Christian churches dominate moral and social life, espousing values that most of the colonising countries banished more than a half century ago, during a period of rapid secularisation. I will return to this post-colonial legacy but other factors are worth noting first.

Sub-Saharan African countries have had a relatively short period of experience with governance and economic dynamics. Prior to colonisation, which began around 1800, the peoples of Africa lived in village communities with tribal structures. During 100 years of colonial rule, the colonising countries imposed Western-style country boundaries, bundling into a single political entity dozens of previously existing tribal communities. Consequently, independence, starting in the 1950s, left sub-Saharan countries with irrational borders, borders within which the only unifying purpose was ending colonial rule; whether the borders were even rational for the colonial powers is debatable but at least they served to define which country controlled which territory, peoples and, most importantly, resources. Transformed by colonial rule, the people could not simply go back to tribal-based villages, although, as the eruption of violence in Kenya in December 2007, after a deeply flawed election, demonstrated, tribal animosities and suspicions are covered by the thinnest civil veneer. Other regions currently described as 'low- and middle-income economies' (China, India and South American countries), although different from each other, all had a much longer history of some form of large-scale governance: from the impressive and large Aztec social structure to the Chinese imperial dynasties. Not so with sub-Saharan African countries. Independence, which, of course, was inevitable and appropriate, resulted, for them, in being cut adrift, rudderless. One consequence is an ongoing dependence on rich-country aid; others are political and administrative corruption, tribalism, electoral fraud and outright internecine violence, which, collectively, grind all hope of a better future into dust.

In addition, malaria and HIV/AIDS have ravaged sub-Saharan Africa. Although much attention has been paid to HIV/AIDS, malaria still kills and weakens more people than any other disease. But the situation is more tragic than the HIV/AIDs or malaria numbers convey. Aid from rich countries is poorly coordinated, competitive and driven by questionable goals, goals that even if well intentioned are often narrow and developed with an eye to a rich-country audience – donors, the electorate and the like. Consequently, billions of dollars have been spent on ameliorating the effects of HIV/AIDS by making, for example, antiretroviral medicines available while food and clean water remain unavailable, and other diseases such as tuberculosis go undiagnosed and untreated. It is completely unreasonable to expect those who are struggling just to stay alive, frequently unsuccessfully, to have the physical, mental and emotional energy to pay any attention to governance, education, family planning and so on.

In many ways, all these factors are part of a post-colonial legacy but it is the ongoing economic dependency of sub-Saharan African countries on rich countries that shackles them the most, and extends colonial control into the present. When I gave a lecture on GM crops to biology faculty and students at the University of Nairobi in 2008, a student commented that he agreed that Kenya would benefit from GM agriculture but then asked, 'What can we do; we are a poor country dependent on trade with Europe whose internal subsidies render our products uncompetitive and whose opposition to GM agriculture is imposed upon us through trade restrictions and NGO propaganda?' What, indeed, can they do? The EU and its member countries wring their hands over the plight of the peoples of African countries but then shackle them economically, assuaging any guilt by generous handouts.

As an aside, those handouts are themselves a curse, a curse Africans accept out of necessity. Biting the hand that feeds you is not usually prudent and so the chains of dependency are seldom thrown off. In her provocative but exceptionally insightful book, *Dead Aid: Why Aid Is Not Working and How There Is a Better Way for Africa*, Dambisa Moyo (2009) deftly outlines the damage of aid to the social, economic and political fabric of African countries. It undermines, for example, the development of functioning markets and makes African-country governments focus more on satisfying the interests of donor groups and countries than the interests of their own citizens. Moyo has experienced this first-hand and has also intellectually examined the impact of aid. She was born and raised in Lusaka, Zambia. She has an MA from Harvard University and a PhD in economics from Oxford University.

The most depressing element in all of this is that while Europeans (and, to be fair, many American and Canadian NGOs) have been lecturing African countries on the evils of GM agriculture, the EU has been steadily dismantling its GM-agriculture restrictions. Recall the transformation, of late, of the precautionary principle (see Section 3.4 above). Also, in 2009, the EU approved several strains of GM maize, and in 2010 a German-developed GM potato was approved for cultivation. Maize is a food staple in many African countries, and

GM maize strains now approved by the EU could benefit small-scale African farmers, but reversing the damage of years of anti-GM propaganda will be a slow process. In addition, it is unlikely that the same level of effort and funding, as was expended in fomenting opposition to GM, will be directed to reversing the now widespread opposition of Africans. Indeed, many NGOs still vigorously oppose GM agriculture and will not be allies in bringing the EU's newfound support to Africans.

This is a rather depressing picture, which focuses on the past and present. Important as it is to understand where we are and how we got here, it is vastly more important, and urgent, to focus on where to go from here. Whatever prescriptions are offered – and these will be many and divergent – all should rest firmly on the recognition of the sovereignty of African nations. Non-Africans may be useful in assisting with the crafting of solutions, but those solutions must be freely adopted by Africans and shaped to their own needs; they must provide tangible, sustainable, long-term benefits to them. In that spirit, African nations must insist, and rich nations acquiesce, that biotechnological research and development, and production and distribution, be part of the fabric of their economies – with research and development being done in their universities and industrial laboratories, and products being produced in their manufacturing plants. Much will have to change in rich nations and in African nations for this to happen.

As a first step, African nations should turn their gaze from Europe, and even North America, and explore the use of GM crops in China, India and South Africa. The picture they present is far from a Utopia – there are still many challenges – but they are making progress. A viable first step would be investigating GM cotton farming in India; cotton is not a food, so a number of apprehensions relating to food (misapprehensions, in my view, as argued above) can be set aside. There are lessons to be learned from the Indian experience (negative and positive) about market forces, incentives, regulations, legal frameworks, taxation, and social benefits and costs. There are lessons to be learned about dramatic increases in profitability, about the transformation of the quality and quantity of cotton by using GM seeds – even on smallholder farms of three or so acres. These are tangible benefits to be found in their experience. As always, there are also social and individual costs, but care must be taken to ensure that a presumed cost can be substantiated. Recently, there have been media and NGO pronouncements, for example, on a connection between farmer suicides and the introduction of *Bt* cotton agriculture in

India. Research, however, has failed to substantiate this connection (Gruère *et al.*, 2008). The causes of suicide in India are complex and GM agriculture appears not to be among the major components. This is an exploration that African nations can begin now, and by so doing they will be taking hold of their destiny and shedding the continuing post-colonial influences of rich nations.

This has to be the long-term strategy; anything else will continue the economic dependence. This will not happen quickly, however, and without some immediate actions – actions consistent with the long-term goal – Africans will continue to die from a lack of nutritious food and clean water. Immediate actions must be focused on increasing local food production. This will require the use of high-yielding crops from quality-controlled seeds – seeds with guaranteed high germination rates and tolerance to local environments. Whether these are from hybrid technology, selective breeding technology or GM technology should be irrelevant. It will also require other inputs such as fertiliser, water and some form of pest control.

Keeping seeds from year to year – no matter how romantic and promoting of self-sufficiency it may seem – compromises the success of increasing yields. As set out in earlier chapters, for good economic and agricultural-practice reasons, farmers in rich counties overwhelmingly purchase quality and trait-true guaranteed seeds from seed companies every year. Increasingly, farmers in low- and middle-income countries like China and India are obtaining seeds from companies on a yearly basis and for the same reason as those in developed countries. Most farmers in African countries – especially small-scale farmers, which constitute the vast majority – do not have the resources to purchase seeds annually and have limited access to such seeds. By default, they retain seeds from one season to plant in the next. This is often romanticised in rich countries as the simpler way of life, living close to the land, self-sufficiency, living in harmony with nature, and avoiding the wicked, iron grip of seed companies, chemical companies and the like. The reality is starkly different. Small-scale African farmers practise this medieval-style agriculture out of necessity and, like peasants in medieval Europe, their days are consumed with staying alive; all too often, they lose that struggle. Hence, to raise this **necessity** to a **virtue** (as some have done) would be simply a bad joke were it not for the fact that the lives and quality of life of millions of Africans are at stake. What African farmers need is access to quality-controlled and trait-guaranteed seeds on a yearly basis, along with other inputs. What are needed are programmes to enable poor African farmers to obtain high-quality seed with traits beneficial to them in their environment.

There is no shortage of innovative and exciting ideas. One that has shown signs of promise in Malawi (see IRIN, 2008) involves companies initially providing free seeds and other appropriate inputs like fertiliser; as a farmer begins to realise a profit above a defensible threshold, a contribution to the purchasing of seed and inputs begins. The contribution will be well below market rates for a long time. If the farmer becomes as successful as farmers in developed countries, as some in China and India are becoming, the contribution will rise to market rates. Many rich-country companies are now working with local African governments on implementing schemes such as this. Of course, this is not charity and the long-term future of African countries depends on it not being charity. As Dambisa Moyo (mentioned above) demonstrates, charity (handouts, foreign aid and so on) is at best a short-term solution to a crisis.

That there is currently a crisis is beyond doubt, which is why virtually no one, not even critics of aid such as Moyo, is suggesting that charity cease. Charity, however, is not a long-run solution. Functioning markets, effective regulatory structures must be developed. These arise from commercial activity. Confidence in Adam Smith's invisible hand of the market was shattered more than a century ago; regulating markets is an essential element of an economic system, but first there have to be markets to regulate. When a company like Syngenta provides seeds and inputs free to African farmers, not as charity, but as a method of developing, in the long-term, a customer base, this ought not to be viewed as a malevolent manipulation of Africans. Rather, it is fostering the creation of markets; Syngenta's self-interest (to call it what it is) results in fostering the achievement of the very goals that African countries need to realise in order to break the bondage of dependence on aid (charity). As with every human social system, this will not usher in a Utopia. Human social systems are messy, complicated, ever changing and prone to excesses. As Winston Churchill commented in a speech in the British House of Commons in 1947, 'Democracy is the worst form of government, except for all those other forms that have been tried from time to time.' An apt twist on this is, 'A regulated capitalist market economy is the worst economic system, except for all those other systems that have been tried from time to time.' Individuals are driven, to greater and lesser extents and with some apparent exceptions, by greed and self-interest. To pretend otherwise is a recipe for failure in developing and maintaining a social system.

To sharpen this point a bit, consider an experience I had several years ago in Kenya. I visited the Unilever tea plantation at Kericho, which is a municipality in the province of Rift Valley in the western region of Kenya. There is also a province of Western Kenya, the capital of which is Kakamega. I am using the designation 'western Kenya' (lower case on 'western') for the region of Kenya west of Nairobi. Unilever has a massive programme of reforestation in western Kenya, as well as the best health-care facilities, HIV/AIDS education and prevention programmes, and provision of food and water. These programmes are not motivated, principally, by environmentalist sentiments or humanitarian impulses. Of course, some people involved will be motivated, in part, by such impulses but there is a major dose of self-interest driving all these programmes. Take the health, food and education programmes. A skilled reliable workforce is essential to a large operation like a tea plantation. Workers compromised by hunger or poor health – especially HIV/AIDS and malaria – do not function at full capacity, and many will die of starvation or disease. Prevention is preferable to constant turnover of workers or compromised functioning of workers. Hence, providing high-quality health care, health education, housing and food is a cost-effective business decision. The environmental reforestation programme is, at its core, similarly a self-interested business decision. Tea requires reliable, predictable rainfall; deforestation has threatened that reliability and predictability. Moreover, Unilever produces its own hydroelectricity. The Kenyan electrical grid is unreliable and the electricity very expensive. Producing electricity by water-driven turbines requires reliable water flow in the rivers. Deforestation has negatively affected those flows. Hence, the compelling business case supports massive reforestation.

Unilever is doing what businesses do; it is making business decisions based on economic self-interest (the self-interest, largely, of executives and shareholders). The result is that Africans who would otherwise be ravaged by disease and starvation win and the environment wins. However much this offends the ideological commitments of some in rich countries, it gets the job done, and, frankly, those with lofty ideals have made little progress in achieving the same goals. The reality is that we are all in this together, whether it is environmental remediation in rich nations or poor ones, or promoting health and adequate nutritious food in rich countries or poor ones. Achieving these always requires a mixed (pluralistic) strategy. An NGO that is advocating some programme of environmental remediation enhances the chances of success if they can co-opt governments, citizens and businesses. To have any one of those sectors pulling in the opposite direction always makes achieving the goal much harder and often dooms it. Co-opting businesses nearly always requires finding the hook of economic self-interest, since acting against that economic self-interest imperils the company's future existence.

The thrust of this line of reasoning is clear. African countries need effectively regulated, strong markets. This requires viable businesses operating in those countries. Viable businesses are viable because they act in their economic self-interest – part of that self-interest resides in satisfying consumer desires and choices, and, of course, in shaping them as well. To cast aspersions on companies that do business in African countries because they are motivated by profit, is to be Polyannaish in the extreme. It also condemns Africans to poverty and continued dependence on rich-country charity, and, therefore, rich-country domination. It is perhaps too cynical to suggest that some NGOs and some rich-country governments want to perpetrate that dependence and domination, but there is, nonetheless, a grain of truth to be teased out of that position. There are NGOs whose *raison d'être* would evaporate if their goals in Africa were accomplished. It is unlikely that these NGOs allow this fact to dominate their thinking and actions, but it is also hard to imagine that it does not lurk somewhere within conscious and unconscious corners of their thoughts. It is also hard to imagine that some anti-business (and anti-other-NGO) rhetoric does not reflect their own competitive impulse driven by self-interest. Some supporters might believe that employees of the NGO they support do not share the same worries about losing their employment as auto company employees, but that belief will rarely be consistent with reality.

I have emphasised plant agriculture to this point for two reasons. First, gains in the quality and quantity of food can be realised more quickly. Second, large-animal agriculture is more precarious and less efficient; less efficient because producing a unit of consumable meat requires many times that unit in plant-derived food and water to raise the animal. Nonetheless, most small-scale African village farms involve mixed agriculture, and there are clear advantages to carefully encouraging mixed agriculture. Milk, from goats, sheep or cows, is a reasonably reliable source of nutrients. Meat from animals provides high-quality protein; to equal meat and milk as a source of protein, most foods from plant sources require dietary complementarities: some mix of grains, beans and nuts, for example. For the reasons set out in Section 2.1, enhancing the desirable traits of farm animals, to this point, has not involved molecular genetic manipulation. Improvements have been achieved mostly by selective breeding, especially by selecting mutations or rarely expressed allelic combinations that confer a beneficial trait – greater milk production, better wool and the like. What African farmers need is access to animals with these trait improvements and access to veterinary services, including artificial insemination, which allow strict selective breeding control. There is, of course, a lot of biotechnology involved in trait enhancement, but it is not molecular and, therefore, opposition is more muted.

A number of programmes designed to provide enhanced-trait animals to African farmers are in place. Heifer International has been very successful in providing animals. In 2010, its website indicated that US$500 would provide a heifer, US$120 a sheep or goat and so on. I have had first-hand experience with the transformation that provision of such animals can bring about. In recognition of the consulting work I have done for Monsanto, a gravid (pregnant) heifer was donated to a local Kenyan NGO, the Rural Outreach Program (ROP), with which I have worked. Monsanto, along with many other companies, has funded many heifer projects. My wife, Jennifer McShane, and I (*wazungu* – 'white people' (plural) – in Swahili) were in the village of a group to which a cow was provided (in the Butere region of western Kenya) the day after the dairy cow arrived. ROP organises women in villages into groups (something like a collective). The group receiving this cow is called *wakulima* (the Swahili word for 'farmer'), and the group chose the member, Evelyn, who would receive the cow. For its part, the group must have built a shelter and planted enough fodder to support the cow before the cow is delivered.

Since the cow was pregnant when it arrived, it shortly gave birth to a calf and began fully lactating. If, as in this case, the calf is female another member of the group receives it; it will be artificially inseminated as, again, will the original cow. If the calf is male, it will be raised until weaning and then sold, with the proceeds supporting other aspects of the group's agricultural endeavours. The cow, donated as a gift to me and which they named Paula, gave birth to a female, which they named Jennifer after my wife, who also works with ROP. It then gave birth to another female, which they named Prof. after the founder of ROP, Hon. Prof. Ruth Oniang'o, who for many years was a professor of nutrition at Jomo Kenyatta University. The transformation this donated cow (donated in 2005) brought about is captured well by my wife in a piece she wrote for the ROP newsletter, after our 2008 visit to the village and the group:

> The next phase of the trip was the much anticipated [return] visit to western Kenya. We flew from Nairobi to Kisumu, on the shores of Lake Victoria, a 40 minute flight. Our first few nights were spent at Golf hotel in Kakamega where I was able to indulge in some great bird watching.
>
> ...
>
> These various groups [ROP women's groups] are well organized with chairpeople, treasurers and secretaries all giving reports. Accountability and commitment are essential for ongoing ROP involvement. Among the projects we visited were those focused on dairy cows, sheep and goats, growing indigenous vegetables, making energy efficient cooking stoves, raising bees. At almost every project we were given "tea". After a meal of ugali, (a polenta/porridge type bread which is a staple of the western diet, and shaped like plum pudding), various indigenous vegetables, cooked chicken, chapatis and a few other things, served with soda, I commented that it was the most interesting tea I had ever had. NO tea had been served. Soda is offered as a preferred drink to guests, and since I never drink pop at home, except for the essential ginger ale when the innards ail, it took some slow sipping to at least be polite and also to avoid a second bottle being immediately opened. The hospitality is without measure.
>
> "Our group", as we call them, is thriving. The Wakulima (Swahili for farmer) group greeted us warmly. The cow, Paula, has had two calves, Jennifer and Prof.; Jennifer is now pregnant. The money from milk sales has funded another building for the family compound, the women in the group have pocket money, and the milk itself is given great credit for helping the children improve school performance. It was a special treat for us to see the heads of these women held high, their success giving them improved self esteem and encouraging ongoing pursuit of group education. (the full text can be found at: www.ropkenya.org/index.php?option=com_content&view=article&id=13:jennifer&catid=11:story1&Itemid=11

These programmes, however, are small steps forward; rewarding as it is to be part of them, progress will be extremely slow unless the pace of change is accelerated considerably. There is much that must change within poor countries in order to accelerate the pace of change. Governments, for example, must become less corrupt and more accountable, and they must put in place effective laws and regulatory structures. Rich countries and their institutions have an essential role to play as well. Engaging corporations in tackling the food, water, education and health challenges of poor countries is essential, as is engaging the governments of rich countries. Promoting GM agriculture has to be part of the solution, which requires changing attitudes that years of anti-GM rhetoric have instilled in Africans. NGO involvement is essential since they are the single largest presence in rich countries. Their influence is still enormous and their infrastructure efficient and effective. To this point, NGOs have been divided on GM agriculture, with many of the largest opposing it. Indeed, the opposition of some seems to be to all agricultural technology. As Robert Paarlberg (2008) notes:

> GRAIN, an NGO concerned with agroecology and genetic diversity headquartered in Barcelona, scolded Bill and Melinda Gates that increased fertilizer use might be of any use to Africans and cited a letter sent earlier by more than 600 NGOs to the Director General of FOA [Federation of Agriculture] which said, "if we have learned anything from the failures of the Green Revolution, it is that technological 'advances' in crop genetics for seeds that respond to external inputs go hand in hand with increased socio-economic polarization, rural and urban impoverishment, and greater food insecurity". (GRAIN, 2006, p. 108)

This is a rather sweeping indictment with which there is much to disagree, but it indicates clearly the opposition of much of the NGO sector to 'technological "advances"', advances from which, it is worth noting, rich countries, and hence the members of these NGOs, have benefited and which they currently happily employ – even leaving aside the recent technological advances of GM. It is not that those who embrace agricultural technology are blind to its downsides; it is that they see **both** the benefits and the harms, consider the benefits to outweigh the harms, and advocate identifying and mitigating or managing harms. To refuse to take an antibiotic when one has, almost certainly fatal, bacterial pneumonia because in the past antibiotics have had side effects such as diarrhoea, vertigo and fatigue betrays a complete lack of understanding of risk analysis. Few people (I suspect no rational person) would prefer death to transient diarrhoea, vertigo and fatigue. So, one troubling aspect of the opposition of 600 NGOs to agricultural technology in Africa – in addition to the hypocrisy of enjoying its benefits in rich countries – is that Africans are starving while potential harms are trotted out in support of denying them the benefits of agricultural technology. Moreover, the simplistic analysis of the purported harms, which avoids any mention of benefits and identifies agricultural technology as their sole cause, is irresponsible; to the extent that the Green Revolution had failings, the causes were many and complex, and it is not at all clear that technology was the most important element in the causal matrix.

This opposition of NGOs might be tolerable, even if misguided, were viable alternatives offered that improved the lives of Africans. However, after spending billions of donated dollars and deploying hundreds of thousands of NGO workers, the situation for most Africans – even in reasonably stable countries – has deteriorated. The alternative to introducing agricultural technology into Africa is, at best, an agricultural *status quo* and many influential NGOs seem to advocate precisely that. Again, Paarlberg (2008) captures this well:

> Strong opposition to Green Revolution-style farming in poor countries nonetheless became a central project of Lappe's NGO (named Food First) created in 1975 [i.e. despite the fact that Lappe and Collins (1977), 'were again likening all the developing world to the most land-unequal nations of Latin America, and they ignored evidence already becoming available that farming in the Green Revolution had helped the poor Asian and could help the poor in Africa as well (Thirtle *et al.*, 2003)']. Working primarily as an advocacy think tank, Food First identified "intensive, externally dependent models of production" as a cause of deepening poverty and growing hunger around the world. The alternative to be promoted was "food production for domestic and local markets based on peasant and family farmer diversified and agroecologically based production systems" (Food First 2002). The problem with this alternative is that it is a perfect technical description of the non-productive, science-starved smallholder farming system that operates in most of rural Africa today. What Food First seems to endorse for the African countryside is little different from the impoverished status quo. (p. 105)

There are promising signs of a change in attitude on the part of some of the better-known NGOs operating in Africa. In November 2008, the Earth Institute of Columbia University, and Monsanto hosted a forum in New York, which involved senior representatives of many of the large NGOs working in Africa, and senior executives of companies such as Syngenta and Nestlé. Jeffery Sachs, director of the Earth Institute and of the UN Millennium Project, and Hugh Grant, the president of Monsanto, were extensively involved in the forum; both advocate more agricultural technology for Africa.

I was at that forum and detected a newfound, albeit cautious, openness to agricultural biotechnology for Africa on the part of the NGO representatives. Building on this co-operative interaction of NGOs and businesses is essential. Fostering the glimmer of openness to agricultural biotechnology, including GM technology, that emerged is vital. On these rest the fate of millions of Africans.

Concluding remarks

In the years ahead, we, and the generations that follow us, will face known and currently unknown challenges. Science and technology will always be a part of the solution for most of them. Hence, opposition to science and technology is almost always imprudent. Equally imprudent is an unquestioning embrace of everything flowing from science and technology. In this spirit, this book has emphasised two main themes. First, important, critical issues are always complex and navigating a course to a resolution requires avoiding extremes and avoiding simple 'this way or the wrong way' dichotomies. That, in part, is why opposition to science and technology (one extreme pole) and unquestioning embrace of science and technology (the other extreme pole) are to be avoided. Second, no activity is without risk of harm – even lying in bed. Activities bring benefits and harms or risks of harms; to pretend otherwise is irrational and imprudent. The rational approach to decision-making involves an examination of benefits and harms (both inextricably value-laden) and balancing them through analysis; if engaging in the activity is the resulting choice, mitigation and management of harms to the maximum extent possible is the rational course. This is the rational and prudent approach for individuals and for societies.

Much of the GM agriculture debate has been dominated by portrayals of the harms – some real, many fabricated. When the harms, for which evidence can be adduced, are balanced with the benefits, for which evidence can be adduced, a more accurate determination of rationally defensible courses of action can be made. When GM crops and the foods derived from them are assessed in this way, the benefits, I contend, outweigh the harms. That has been a central message of the book, a message around which much of the argumentation has revolved. More importantly, though, success, for me, will not be measured in terms of the number of people who come to agree with its message but rather in terms of the number of people whose analytical skills have been enhanced. Hence, someone who rejects the central message but does so by using evidence and analytical tools is an instance of success. In the end, dialogue and debate is not about winning but about grappling rationally with complex matters and developing tentative positions, positions that can, and almost certainly will, be refined and transformed over time. That is the intellectually honest and most beneficial (individually and socially) approach to forming positions and making decisions.

Bibliography

Adam, John A. (2009) *A Mathematical Nature Walk*. Princeton University Press.

Ambec, Stefan (2005) *Risk Regulation Related to Resistance Development in Insects*. Department of INRA Social Sciences, Agriculture and Food, Space and Environment Publishing Unit, Institute National de la Recherche Agronomique (INRA).

American Law Institute (1965–79) *Restatement of the Law, Second, Torts 2d*. St. Paul, MN: American Law Institute Publishers.

Ames, B. N., *et al.* (1987) Ranking possible carcinogenic hazards, *Science* **236**: 271–280.

(1990) Dietary pesticides (99.99% all natural), *Proceedings of the National Academy of Sciences of the USA* **87**: 7777–7781.

Ammann, Klaus (2008) Integrated farming: why organic farmers should use transgenic crops, *New Biotechnology* **25**: 101–107.

(2009) Why farming with high tech methods should integrate elements of organic agriculture, *New Biotechnology* **25**: 378–388.

Arrow, Kenneth (1973) Rawls's principle of just savings, *Swedish Journal of Economics* **75**: 323–325.

Avery, D. T. (1998) The hidden dangers in organic food, *American Outlook*, **Fall**: 19–22. www.hudson.org/index.cfm?fuseaction=publication_details&id=1196.

Badgley, Catherine, *et al.* (2007) Organic agriculture and the global food supply, *Renewable Agriculture and Food Systems* **22**(2): 86–108.

Bahlai, C. A., *et al.* (2010) Choosing organic pesticides over synthetic pesticides may not effectively mitigate environmental risk in soybeans, *PLoS ONE* **5**(6): e11250. doi:10.1371/journal.pone.0011250.

Beauchamp, Tom L. and Childress, James F. (2009) *Principles of Biomedical Ethics*, 6th edn. New York: Oxford University Press.

Beers, Mark H. and Berkow, Robert (1999) *The Merck Manual of Diagnosis and Therapy*, 17th edn. Whitehouse Station, NJ: Merck Research Laboratories.

Beije, Brita and Möller, Lennart (1988) 2-Nitrofluorene and related compounds: prevalence and biological effects, *Mutation Research/Reviews in Genetic Toxicology* **196**(2): 177–209.

Bentham, Jeremy (1789) *An Introduction to the Principles of Morals and Legislation*. London: T. Payne.

Bourguet, D., *et al.* (2005) Regulating insect resistance management: the case of non-Bt corn in the U.S., *Journal of Environmental Management* **76**: 210–220.

Bourn, Diane and Prescott, John (2002) A comparison of the nutritional value, sensory qualities, and food safety of organically and conventionally produced foods, *Food Science and Nutrition* **42**: 1–34.

Brigulla, Matthias and Wackernagel, Wilfried (2010) Molecular aspects of gene transfer and foreign DNA acquisition in prokaryotes with regard to safety issues, *Applied Microbiology and Biotechnology* **86**: 1027–1041. doi: 10.1007/s00253-010-2489-3.

Brunk, C. and Coward, H. (2009) *Acceptable Genes? Religious Traditions and Genetically Modified Foods*. Albany, NY: State University of New York Press.

Buckle, Stephen (1993) Natural law, in Peter Singer (ed.), *A Companion to Ethics*. Oxford: Blackwell, pp. 161–174.

Caduff, Ladina (2002) Growth hormones and beyond, Working Paper 8-2000, Center for International Studies: Eidgenössische Technische Hochschule Zürich (Swiss Federal Institute of Technology, Zurich).

Castle, David (2003) The moral significance of agricultural biotechnology, *Studies in History and Philosophy of Biological and Biomedical Sciences* **34**: 713–722.

(2006) The balance between expertise and authority in citizen engagement about new biotechnology, *Techne* **9**: 1–13.

Castle, David and Culver, K. (2006) Public engagement, public consultation, innovation and the market, *Integrated Assessment Journal* **6**: 137–152.

Castle, David, Finlay K. and Clark, S. (2005) Proactive consumer consultation: the effect of information provision on response to transgenic animals, *Journal of Public Affairs* **5**: 200–216.

Castle, David, *et al.* (2006) *Science, Society and the Supermarket: The Opportunities and Ethical Challenges of Nutritional Genomics*. Hoboken, NJ: Wiley.

Chandrasekharan, N. V. and Simmons, Daniel L. (2004) The cyclooxygenases, *Genome Biology* **5**: 241. doi:10.1186/gb-2004-5-9-241.

Cho, Y. R. and Palmer, J. D. (1999) Multiple acquisitions via horizontal transfer of a group I intron in the mitochondrial cox1 gene during evolution of the Araceae family, *Molecular Biology and Evolution* **16**: 1155–1165.

Cho, Y., *et al.* (1998) Explosive invasion of plant mitochondria by a group I intron, *Proceedings of the National Academy of Sciences of the USA* **95**: 14244–14249.

Christiansen, F. B. (1978) Genetics of *Zoarces* populations. X. Selection component analysis of the *EstIII* polymorphism using samples of successive cohorts, *Hereditas* **87**: 129–150.

Church of England (2010) Special agenda IV. Diocesan synod motions. Compatibility of science and Christian belief. A background paper from the Diocese of Manchester, GS 1772A.

Clarke, R. J. and Macrae, R. (eds) (1988) *Coffee*, vols. I–III. New York: Elsevier.

Colborn, T., Dumanoski, D. and Myers, J. P. (1996) *Our Stolen Future: Are We Threatening Our Fertility, Intelligence, and Survival?: A Scientific Detective Story*. London: Penguin Books.

Commission of the European Community (2000) *Communication from the Commission on the Precautionary Principle*. Brussels, 2.2.2000 COM(2000) 1 final.

Crespo, A. L. B., *et al.* (2009) On-plant survival and inheritance of resistance to Cry1Ab toxin from *Bacillus thuringiensis* in a field-derived strain of European corn borer, *Ostrinia nubilalis*, *Pest Management Science* **65**: 1071–1081.

Crow, Ernest W. and Crow, James F. (2002) 100 years ago: Walter Sutton and the chromosome theory of heredity, *Genetics* **160**: 1–4.

Cullis, Christopher A. (2004) *Plant Genomics and Proteomics*. Hoboken, NJ: Wiley.

Daniels, N. (1975) (ed.) *Reading Rawls*. Oxford: Blackwell.

Darwin, Charles (1859) *On the Origin of Species*. London: John Murray.

(1876) *The Effects of Cross and Self Fertilisation in the Vegetable Kingdom*. London: J. Murray.

Deplanque, Catherine (2004) Origins and Impact of the French Civil Code, Association Française pour l'Histoire de la Justice (20 July).

Diamond, Jared (2002) Evolution, consequences and future of plant and animal domestication, *Nature* **418**: 700–707.

Dieckmann, W. J., *et al.* (1953) Does the administration of diethylstilbestrol during pregnancy have therapeutic value?, *American Journal of Obstetrics and Gynecology* **66**: 1062–1081.

Dodds, Edward Charles, *et al.* (1938) Estrogenic activity of certain synthetic compounds, *Nature* **141**: 247–248.

Eckhardt, R. B. (2001) Genetic research and nutritional individuality, *Journal of Nutrition* **131**: 336S–339S.

Elias, Thomas S. and Dykeman, Peter A. (1990) *Edible Wild Plants: A North American Guide*. New York: Sterling.

Ellstrand, Norman C. (2001) When transgenes wander, should we worry?, *Plant Physiology* **125**: 1543–1545.

Environmental Protection Agency (EPA) (2001) *Biopesticides Registration Action Document*. Section D: *Bt Plant-Incorporated Protectants*. Washington, DC: EPA.

Ewen, Stanley W. B. and Pusztai, Arpad (1999) Effects of diets containing genetically modified potatoes expressing *Galanthus nivalis* lectin on rat small intestine, *The Lancet* **354**: 1353–1354.

Eyestone, W. H. (1998) Techniques for the production of transgenic livestock, in A. J. Clark (ed.), *Animal Breeding: Technology for the 21st Century*. Amsterdam: Harwood Academic Publishers, pp. 167–182.

Fay, Maryanne and Bierbaum, Rosina (2010) *The World Development Report 2010: Development and Climate Change*. World Bank.

Feinberg, Joel (1973) *Social Philosophy*. Englewood Cliffs, NJ: Prentice-Hall.

Fisher, Ronald A. (1930) *The Genetical Theory of Natural Selection*. Oxford: Clarendon Press.

Fisher, R. A. (1935) *The Design of Experiments*. Edinburgh: Oliver & Boyd.

Food Standards Agency (2000) Position Paper: Food Standards Agency View on Organic Foods, 23 August.

(2010) Position Paper: Food Standards Agency View on Organic Foods.

Forge, Frédéric (2004) Organic farming in Canada: an overview, *Parliamentary Information and Research Service/Service d'information et de recherche parlementaires* (Science and Technology Division), *PRB* 00–29E.

Formation of Mutagens During Cooking (Special Issue) (1986) *Environmental Health Perspectives*, **67**, 3–157.

Freud, Sigmund (1979) *Introductory Lectures on Psychoanalysis*, transl. and ed. James Strachey. New York: Liveright Publishing.

Furihata, C. and Matsushima, T. (1986) Mutagens and carcinogens in foods, *Annual Review of Nutrition* **6**: 67–94.

Garcia-Robles, I., *et al.* (2001) Mode of action of *Bacillus thuringiensis* PS86Q3 strain in hymenopteran forest pests, *Insect Biochemistry and Molecular Biology* **31**: 849–856.

Gardon, Anne (1998) *The Wild Food Gourmet: Fresh and Savory Food from Nature*. Willowdale, ON: Firefly Books.

Gilbert-López, Bienvenida, *et al.* (2009) Sample treatment and determination of pesticide residues in fatty vegetable matrices: a review, *Talanta* **79**: 109–128.

Glare, T. R., and O'Callaghan, M. (2000) *Bacillus thuringiensis: Biology, Ecology and Safety*. Chichester: Wiley.

Gold, L. S., *et al* (1984) *A Carcinogenic Potency Database of the Standardized Results of Animal Bioassays*, in *Environmental Health Perspectives* **58**: 9–319.

(1986) *Chronological Supplement to the Carcinogenic Potency Database: Standardized Results of Animal Bioassays Published Through December 1982*, in *Environmental Health Perspectives* **67**: 161–200.

(1987) *Second Chronological Supplement to the Carcinogenic Potency Database: Standardized Results of Animal Bioassays Published Through December 1986*, in *Environmental Health Perspectives* **74**: 237–329.

Goulding, K. W. T. and Trewavas, A. J. (2009) Can organic agriculture feed the world?, *AgBioView*, 23 June. www.agbioworld.org.

GRAIN (2006) Another silver bullet for Africa? Bill Gates to resurrect the Rockefeller Foundation's decaying green revolution. www.grain.org/articles/?id-19.

Gruère, Guillaume P., Mehta-Bhatt, Purvi and Sengupta, Debdatta (2008) Bt cotton and farmer suicides in India: reviewing the evidence, International Food Policy Research Institute: Discussion Paper 00808.

Gunderson, E. L. (1988) FDA Total Diet Study, April 1982–April 1984, dietary intakes of pesticides, selected elements, and other chemicals, *Journal of the Association of Official Analytical Chemists* **71**: 1200–1209.

Haldane, J. B. S. (1924–32) A mathematical theory of natural and artificial selection, (9 parts) *Transactions and Proceedings of the Cambridge Philosophical Society*.

(1932) *The Causes of Evolution*. London: Longmans, Green.

Hallauer, Arnel R. (1978) Potential of exotic germplasm for maize improvement, in Walden (ed.), *Maize Breeding and Genetics*, pp. 229–247.

Hardy, G. H. (1908) Mendelian proportions in a mixed population, *Science* **28**: 49–50.

Hare, R. M. (1952) *The Language of Morals*. Oxford: Clarendon Press.

(1964) *Freedom and Reason*. Oxford: Clarendon Press.

Harsanyi, John (1975) Can the maximin principle serve as the basis for morality? A critique of Rawls's theory, *American Political Science Review* **69**: 594–606.

Henderson, Robert K. (2000) *The Neighborhood Forager: A Guide for the Wild Food Gourmet*. White River Junction, VT: Chelsea Green Publishing.

Hobbes, Thomas [1651] *Leviathan*. London. For a recent reissue, see J. C. A. Gaskin, *Leviathan* (edited with introduction). New York: Oxford University Press, 2008.

House of Commons (UK) Agricultural Committee (2001) Agriculture – Second Report. www.parliament.the-stationery-office.co.uk/pa/cm200001/cmselect/cmagric/149/14902.htm.

Howson, Colin (2000) *Hume's Problem: Induction and the Justification of Belief*. Oxford University Press.

Howson, Colin and Urbach, Peter (1989) *Scientific Reasoning: The Bayesian Approach*. La Salle, IL: Open Court.

HRH The Prince of Wales (2000) A royal view, in Chris Patten *et al.* (eds), *Respect for the Earth: Sustainable Development*. Reith Lectures 2000. London: Profile – BBC.

Hume, David (1739) *A Treatise of Human Nature*. London: John Noon [reprinted 2000, David Fate Norton and Mary J. Norton (eds), New York: Oxford University Press, 2000].

(1777) *An Enquiry Concerning Human Understanding* [reprinted, Tom L. Beauchamp (ed.), *An Enquiry Concerning Human Understanding: A Critical Edition*. New York: Oxford University Press, 2000].

(1779) *Dialogues Concerning Natural Religion* [reprinted, Dorothy Coleman (ed.), *Dialogues Concerning Natural Religion and Other Writings*. Cambridge University Press, 2007].

Illich, Ivan (1977) *Limits to Medicine*. Harmondsworth: Penguin Books.

IRIN: *Humanitarian News and Analysis* (2008) Malawi: subsidising agriculture is not enough. www.irinnews.org/Report.aspx?ReportId=76591.

John Paul II (1988) Letter to the Reverened George V. Coyne, S. J., Director of the Vatican Observatory, *L'Osservatore Romano* (Weekly edition in English), **21**(46), 14 November. www.its.caltech.edu/~nmcenter/sci-cp/sci-coyne.html.

Jowett, Benjamin (1931) *The Dialogues of Plato*, 3rd edn. Translated into English with analyses and introduction by B. Jowett. London: Oxford University Press.

Kagan, S. (1998) *Normative Ethics*. Boulder, CO: Westview Press.

Kalaitzandonakes, Nicholas (ed.) (2003) *The Economic and Environmental Impacts of Agriotech: A Global Perspective*. New York: Kluwer.

Kant, Immanuel (2002) *Groundwork for the Metaphysics of Morals*, transl. and ed. Allen W. Wood. New Haven, CT: Yale University Press.

Kilham, P. and Hecky, R. E. (1973) Fluoride: geochemical and ecological significance in East Africa waters and sediments, *Limnology and Oceanography* **18**: 932–945.

Kjellevold Malde, Marian, *et al.* (1997) Fluoride content in selected food items from five areas in East Africa, *Journal of Food Composition and Analysis* **10**: 233–245.

Korsgaard, Christine M. (1992) Kant, Immanuel, in Lawrence C. Becker and Charlotte B. Becker (eds), *Encyclopedia of Ethics*. New York: Garland Publishing, pp. 664–674.

Kravitz, Richard L., *et al.* (2004) Evidence-based medicine, heterogeneity of treatment effects, and the trouble with averages, *Milbank Quarterly* **82**: 661–687.

Kumpulainen, J. (2001) Organic and conventional grown foodstuffs: nutritional and toxicological quality comparisons, *Proceedings of the International Fertiliser Society* **472**: 1–20.

Lachmann, Peter (1999) Letter, *The Lancet* **354**: 1726 (see also Ewen and Pusztai response, *The Lancet* **354**: 1726–1727).

Lappe, Frances Moore and Collins, Joseph (1977) *Food First: Beyond the Myth of Scarcity*. Boston, MA: Houghton Mifflin.

Le Fanu, James (2000) *The Rise and Fall of Modern Medicine*. New York: Carroll & Graf.

Lercher, M. J. and Pal, C. (2008) Integration of horizontal transferred genes into regulatory interaction networks takes many million years, *Molecular Biology and Evolution* **25**: 559–567.

Liddell, H. G. (1966). *A lexicon abridged from Liddell and Scott's Greek-English Lexicon*. Oxford: Clarendon Press.

Lind, Niels C., *et al.* (1993) *Health and Safety Policies: Guiding Principles for Risk Management.* Waterloo, ON: Institute for Risk Research (Report 93–1, Joint Committee on Health and Safety of the Royal Society of Canada and the Canadian Academy of Engineering).

Locke, John (1690, 1691) *Two Treatises of Government.* Reprint: Cambridge University Press, 1960.

Losey, John E., Raynor, Linda S. and Carter, Maureen E. (1999) Transgenic pollen harms monarch larvae, *Nature* **399**: 214.

Lynch, Michael and Walsh, Bruce (1998) *Genetics and Analysis of Quantitative Traits.* Sunderland, MA: Sinauer.

Maarse, H. and Visscher, C. A. (eds) (1989) *Volatile Compounds in Foods.* Zeist, The Netherlands: CIVO-TNO.

McGee, Harold (1997) *On Food and Cooking: The Science and Lore of the Kitchen*, rev. edn. New York: Simon & Schuster.

Magkos, Faidon, Arvaniti, Fotini and Zampelas, Antonis (2006) Organic food: buying more safety or just peace of mind? A critical review of the literature, *Critical Reviews in Food Science and Nutrition* **46**: 23–56.

Malthus, Thomas (1798) *An Essay on the Principle of Population.* London: J. Johnson.

Mangelsdorf, Paul C. (1974) *Corn: Its Origin, Evolution and Improvement.* Cambridge, MA: Harvard University Press.

Marroquin, L. D. *et al.* (2000) *Bacillus thuringiensis* (*Bt*) toxin susceptibility and isolation of resistance mutants in the nematode *Caenorhabditis elegans*, *Genetics* **155**: 1693–1699.

Marvier, M., *et al.* (2007) A meta-analysis of effects of *Bt* cotton and maize on nontarget invertebrates, *Science* **316**: 1475–1477.

May, R. M. (1999) Genetically modified foods: facts, worries, policies and public confidence. London: UK Office of Science and Technology.

Mazoyer, Marcel and Roundart, Laurence (2006) *A History of World Agriculture from the Neolithic Age to the Current Crisis* (translated by James H. Membrez). New York: Monthly Review Press, 94–95.

Menard, Céline, *et al.* (2008) Relevance of integrating agricultural practices in pesticide dietary intake indicator, *Food and Chemical Toxicology* **46**: 3240–3253.

Mendel, Gregor (1865) Versuche über Pfanzenhybriden, *Verhandlungen des Naturforschenden Vereins Brünn* **4**.

Metcalf, R. L. (1989) Insect resistance to insecticides, *Pesticide Science* **26**: 333–358.

Mill, John Stuart (1859) *On Liberty*. London: John W. Parker and Son.

(1863) *Utilitarianism*. London: Parker, Son, and Bourn.

Moore, G. E. (1903) *Principia Ethica*. Cambridge University Press.

Monsanto Company (2010) Observations on competition in the U.S. seed industry. www.monsanto.com/choice_in_agriculture/monsanto_submission_doj.aspx.

Moyo, Dambisa (2009) *Dead Aid: Why Aid Is Not Working and How There Is a Better Way for Africa*. New York: Farrar, Straus and Giroux.

Munkvold, G. P. and Hellmich, R. L. (2000) Genetically modified, insect resistant maize: implications for management of ear and stalk diseases, *Plant Health Progress* (online) doi:10.1094/PHP-2000–0912–01-RV.

Nakamura, Yoji, *et al.* (2004) Biased biological functions of horizontally transferred genes in prokaryotic genomes, *Nature Genetics* **36**: 760–766.

Nathwani, J. S., *et al.* (1997) *Affordable Safety by Choice: The Life Quality Model*. Waterloo, ON: Institute for Risk Research.

National Research Council (1989) *Diet and Health: Implications for Reducing Chronic Disease Risk*. Washington, DC: National Academy Press.

Navarre, W. W., *et al.* (2007) Silencing of xenogeneic DNA by H-NS – facilitation of lateral gene transfer in bacteria by a defense system that recognizes foreign DNA, *Genes and Development* **21**: 1456–1471.

Nicolis, Grégoire and Prigogine, Ilya (1989) *Exploring Complexity: An Introduction*. New York: W. H. Freeman.

Nigg, H. N., *et al.* (1990) Exposure to pesticides, in Scott R. Baker and Chris F. Wilkinson (eds), *The Effects of Pesticides on Human Health*. Advances in Modern Environmental Toxicology, vol. XVIII. Princeton, NJ: Princeton Scientific Publishing, pp. 35–130.

Organisation for Economic Co-operation and Development (2009) OECD. Stat Extracts. http://stats.oecd.org/WBOS/index.aspx.

Paarlberg, Robert (2008) *Starved for Science: How Biotechnology Is Being Kept out of Africa*. Cambridge, MA: Harvard University Press.

Paley, William (1809 [1802]) *Natural Theology; or, Evidences of the Existence and Attributes of the Deity Collected from the Appearances of Nature*, 12th edn. London: J. Faulder.

Paton, H. J. (1971) *The Categorical Imperative: A Study in Kant's Moral Philosophy*. Philadelphia: University of Pennsylvania Press.

Pauling, Linus, *et al.* (1951) The structure of proteins: two hydrogen-bonded helical configurations of the polypeptide chain, *Proceedings of the National Academy of Sciences of the USA* **37**: 205–211.

Perry, Mark (2008) From Pasteur to Monsanto: approaches to patenting life in Canada, in Ysolde Gendreau (ed.), *An Emerging Intellectual Property Paradigm – Perspectives from Canada*. Northampton, MA: Edward Elgar Publishing, pp. 67–80.

Peterson, John (1989) Hormones, heifers and high politics – biotechnology and the common agricultural police, *Public Administration* **67**: 455–471.

Pigott, Craig R. and Ellar, David J. (2007) Role of receptors in *Bacillus thuringiensis* crystal toxin activity, *Microbiology and Molecular Biology Reviews* **71**(2): 255–281.

Population Reference Bureau (2008). www.prb.org/Publications/Datasheets/2008/2008wpds.aspx.

Priest, Graham (1986) Contradiction, belief, and rationality, *Proceedings of the Aristotelian Society* **86**: 99–116.

Rawls, John (1971) *A Theory of Justice*. Cambridge, MA: Harvard University Press.

(1993) *Political Liberalism*. New York: Columbia University Press.

(2000) *Lectures on the History of Moral Philosophy*, ed. Barbara Herman. Cambridge, MA: Harvard University Press.

Richardson, Aaron O. and Palmer, Jeffrey D. (2007) Horizontal gene transfer in plants, *Journal of Experimental Botany* **58**: 1–9.

Ridley, Mark (1996) *Evolution*, 2nd edn. Cambridge, MA: Blackwell Science.

Rogers, P. L. (2002). *The Organic Factor: The Whole Food Way to Health and Fitness for Life*. Keperra, Queensland: Ecol Data Pty Limited.

Ronald, Pamela C. and Adamchak, Raoul W. (2008) *Tomorrow's Table: Organic Farming, Genetics and the Future of Food*. New York: Oxford University Press.

Ruse, Michael (1975) Darwin's debt to philosophy: an examination of the influences of John F. W. Herschel and William Whewell on the development of Charles Darwin's theory of evolution, *Studies in History and Philosophy of Science* **6**: 159–181.

(1999) *The Darwinian Revolution: Science Red in Tooth and Claw*, 2nd edn. Chicago University Press.

Ruse, Michael and Castle, David (eds) (2002) *Genetically Modified Foods: Debating Biotechnology*. Amherst, NY: Prometheus Books.

Sachs, Jeffery D. (2005) *The End of Poverty: Economic Possibilities for Our Time*. New York: Penguin Press.

Salsburg, David (1993) The use of statistical methods in the analysis of clinical studies, *Journal of Clinical Epidemiology* **46**: 17–27.

Sears, Mark K., *et al.* (2001) Impact of *Bt* corn pollen on monarch butterfly populations, *Proceedings of the National Academy of Sciences of the USA* **98**: 11937–11942.

Sen, Amartya (1992) *Inequality Reexamined*. Cambridge, MA: Harvard University Press.

Shelley, Mary W. (1818) *Frankenstein* (reprinted numerous times: see Paul J. Hunter (ed.), *Frankenstein*. Norton Critical Edition. New York: W. W. Norton, 1996).

Shiva, Vandana (1997) *Biopiracy: The Plunder of Nature and Knowledge*. Boston, MA: South End Press.

Sikorski, James A. and Gruys, Kenneth J. (1997) Understanding glyphosate's molecular mode of action with EPSP synthase: evidence favoring an allosteric inhibitor model, *Accounts of Chemical Research* **30**(1): 2–8.

Skyrms, Brian (1966) *Choice and Chance: An Introduction to Inductive Logic*. Belmont, CA: Dickenson Publishing.

Smith, Charles (1998) Introduction: current animal breeding, in A. J. Clark (ed.), *Animal Breeding: Technology for the 21st Century*. Amsterdam: Harwood Academic, pp. 1–10.

Spector, Michael (2009) *Denialism: How Irrational Thinking Hinders Scientific Progress, Harms the Planet, and Threatens Our Lives*. New York: Penguin Press.

Starr, Paul (1982) *The Social Transformation of American Medicine*. New York: Basic Books.

Statistics Canada (2003) *Death by Causes, 2000*. Ottawa, ON: Statistics Canada.

(2006) *Causes of Death, 2003*. Ottawa, ON: Statistics Canada.

Steinfeld, Henning, *et al.* (2006) *Livestock's Long Shadow: Environmental Issues and Options*. Rome: Food and Agriculture Organization of the United Nations.

Stofberg, J. and Grundschober, F. (1987) Consumption ratio and food predominance of flavoring materials, *Perfumer and Flavorist* **12**: 27–56.

Sugimura, T. (1988) Successful use of short-term tests for academic purposes: their use in identification of new environmental carcinogens with possible risk for humans, *Mutation Research/Genetic Toxicology* **205**: 33–39.

Sullivan, Walter (1968) A book that couldn't go to Harvard, *New York Times*, 15 February, pp. 1 and 4.

Sutton, Walter S. (1902) On the morphology of the chromosome group in *Brachystola magna*, *Biological Bulletin* **4**: 24–39.

(1903) The chromosomes in heredity, *Biological Bulletin* **4**: 231–251.

Takayama, Shozo, *et al.* (1987) Carcinogenic effects of the simultaneous administration of five heterocyclic amines to F344 rats, *Japanese Journal of Cancer Research* **78**: 1068–1072.

Tape, Cynthia L. and McCourt, Conor D. M. (2003) Supreme Court kills Harvard's mouse, *BioBusiness* **30 April**: 16–19.

Thayer, Samuel (2006) *The Forager's Harvest: A Guide to Identifying, Harvesting, and Preparing Edible Wild Plants*. Ogema, WI: Forager's Harvest.

Thirtle, C., Lin, L. and Piesse, J. (2003) The impact of research-led agricultural productivity growth on poverty reduction in Africa, Asia and Latin America, *World Development* **31**: 1959–1975.

Thompson, Paul (2002) The evolutionary biology of evil, *The Monist* **85**: 239–259.

(2009) History of scientific agriculture: animals, in *Encyclopedia of Life Sciences*. Chichester: Wiley. doi:10.1002/9780470015902.a0020136.

Thompson, Paul (2010) Causality, mathematical models and statistical association: dismantling evidence-based medicine, *Journal of Evaluation in Clinical Practice* **16**: 267–275.

(2011a) Causality, theories and medicine, in Phyllis McKay Illari, Federica Russo and Jon Williamson (eds), *Causality in the Sciences*. Oxford University Press, pp. 25–44.

(2011b) Theories and models in medicine, in Fred Gifford (ed.), *Handbook of the Philosophy of Science. Volume 16: Philosophy of Medicine*, 117–138.

Treon, J. F., Dutra, F. R. and Cleveland, F. P. (1953) Toxicity of 2-ethylhexyl diphenyl phosphate. I. Immediate toxicity and effects of long-term feeding experiments, *American Medical Association Archives of Industrial Hygiene and Occupational Medicine* **8**(2): 170–184.

Trewavas, A. and Stewart, D. (2003) Paradoxical effects of chemicals in the diet on health, *Current Opinion in Plant Biology* **6**: 185–190.

Troyer, A. Forest (2006) Adaptedness and heterosis in corn and mule hybrids, *Crop Science* **46**: 528–543.

Unilever Corporate site (n.d.) Unilever at a glance. www.unilever.com/aboutus/introductiontounilever/unileverataglance/index.aspx.

United States Environmental Protection Agency (n.d.) Consumer Factsheet on glyphosate. www.epa.gov/ogwdw/pdfs/factsheets/soc/glyphosa.pdf.

University of California, San Diego (n.d.) Agriculture. www.bt.ucsd.edu/bt_safety.html

Upshur, Ross E. G. (2005) Looking for rules in a world of exceptions: reflections on evidence-based practice, *Perspectives* **48**: 477–489.

USDA/NASS (2009) Crop Production Historical Track Records. US Department of Agriculture/National Agricultural Statistics Service. http://usda.mannlib.cornell.edu/MannUsda/viewDocumentInfo.do;jsessionid=84F77281F843295CB6CE233C52D5B46F?documentID=1593.

Van Valen, Leigh (1973) A new evolutionary law, *Evolutionary Theory* **1**: 1–30.

Von Wright, Georg Henrik (1960) *A Treatise on Induction and Probability*. Patterson, NJ: Littlefield, Adams & Company.

Walden, David B. (ed.) (1978) *Maize Breeding and Genetics*. New York: Wiley.

Watkins, S. M., Hammock, B. D., Newman, J. W. and German, J. B. (2001) Individual metabolism should guide agriculture toward foods for improved health and nutrition, *American Journal of Clinical Nutrition* **74**: 283–286.

Watson, James D. (1968) *The Double Helix*. New York: Atheneum.

Watson, J. D. and Crick, F. H. C. (1953a) Molecular structure of nucleic acids: a structure for deoxyribose nucleic acid, *Nature* **171**: 737–738.

(1953b) Genetical implications of the structure of deoxyribonuleic acid, *Nature* **171**: 964–967.

Wei, J. Z., *et al.* (2003) *Bacillus thuringiensis* crystal proteins that target nematodes, *Proceedings of the National Academy of Sciences of the USA* **100**: 2760–2765.

Weinberg, Wilhelm (1908) Über den Nachweis der Vererbung beim Menschen, *Jahreshefte des Vereins für vaterländische Naturkunde in Württemberg* **64**: 368–382.

Whewell, William (1840) *Philosophy of the Inductive Sciences*. London: Parker.

Wills, Gary (1978) *Inventing America: Jefferson's Declaration of Independence*. Garden City, NY: Doubleday.

Wilson, Edward Osborne (1975) *Sociobiology: The New Synthesis*. Cambridge, MA: Harvard University Press.

World Bank (2000) World Development Indicators. http://web.worldbank.org/WBSITE/EXTERNAL/DATASTATISTICS/0,contentMDK:21725423~pagePK:64133150~piPK:64133175~theSitePK:239419,00.html.

World Trade Organization (WTO) (1998) *EC Measures Concerning Meat and Meat Products (Hormones)*. 16 January 1998, AB-1997, WT/DS26/AB/R; WT/DS48/AB/R.

World Wildlife Fund (n.d.) www.worldwildlife.org/cci/agriculture.cfm.

Worrall, J. (2002) *What* evidence in evidence-based medicine?, *Philosophy of Science* **69**: S316–S330.

Wright, Sewall (1931) Evolution in Mendelian populations, *Genetics* **16**: 97–159.

Index